Individual and Collective Graph Mining

Principles, Algorithms, and Applications

Synthesis Lectures on Data Mining and Knowledge Discovery

Editor

Jiawei Han, *University of Illinois at Urbana-Champaign*
Lise Getoor, *University of California, Santa Cruz*
Wei Wang, *University of California, Los Angeles*
Johannes Gehrke, *Cornell University*
Robert Grossman, *University of Chicago*

Synthesis Lectures on Data Mining and Knowledge Discovery is edited by Jiawei Han, Lise Getoor, Wei Wang, Johannes Gehrke, and Robert Grossman. The series publishes 50- to 150-page publications on topics pertaining to data mining, web mining, text mining, and knowledge discovery, including tutorials and case studies. Potential topics include: data mining algorithms, innovative data mining applications, data mining systems, mining text, web and semi-structured data, high performance and parallel/distributed data mining, data mining standards, data mining and knowledge discovery framework and process, data mining foundations, mining data streams and sensor data, mining multi-media data, mining social networks and graph data, mining spatial and temporal data, pre-processing and post-processing in data mining, robust and scalable statistical methods, security, privacy, and adversarial data mining, visual data mining, visual analytics, and data visualization.

Individual and Collective Graph Mining: Principles, Algorithms, and Applications
Danai Koutra and Christos Faloutsos
2017

Phrase Mining from Massive Text and Its Applications
Jialu Liu, Jingbo Shang, and Jiawei Han
2017

Exploratory Causal Analysis with Time Series Data
James M. McCracken
2016

Mining Human Mobility in Location-Based Social Networks
Huiji Gao and Huan Liu
2015

Mining Latent Entity Structures
Chi Wang and Jiawei Han
2015

Probabilistic Approaches to Recommendations
Nicola Barbieri, Giuseppe Manco, and Ettore Ritacco
2014

Outlier Detection for Temporal Data
Manish Gupta, Jing Gao, Charu Aggarwal, and Jiawei Han
2014

Provenance Data in Social Media
Geoffrey Barbier, Zhuo Feng, Pritam Gundecha, and Huan Liu
2013

Graph Mining: Laws, Tools, and Case Studies
D. Chakrabarti and C. Faloutsos
2012

Mining Heterogeneous Information Networks: Principles and Methodologies
Yizhou Sun and Jiawei Han
2012

Privacy in Social Networks
Elena Zheleva, Evimaria Terzi, and Lise Getoor
2012

Community Detection and Mining in Social Media
Lei Tang and Huan Liu
2010

Ensemble Methods in Data Mining: Improving Accuracy Through Combining Predictions
Giovanni Seni and John F. Elder
2010

Modeling and Data Mining in Blogosphere
Nitin Agarwal and Huan Liu
2009

Individual and Collective Graph Mining: Principles, Algorithms, and Applications

Danai Koutra and Christos Faloutsos

ISBN: 978-3-031-00783-5   paperback
ISBN: 978-3-031-01911-1   ebook

DOI: 10.1007/978-3-031-01911-1

A Publication in the Springer series
SYNTHESIS LECTURES ON DATA MINING AND KNOWLEDGE DISCOVERY

Lecture #14
Series Editors: Jiawei Han, *University of Illinois at Urbana-Champaign*
　　　　　　　　Lise Getoor, *University of California, Santa Cruz*
　　　　　　　　Wei Wang, *University of California, Los Angeles*
　　　　　　　　Johannes Gehrke, *Cornell University*
　　　　　　　　Robert Grossman, *University of Chicago*
Series ISSN
Print 2151-0067   Electronic 2151-0075

Individual and Collective Graph Mining

Principles, Algorithms, and Applications

Danai Koutra
University of Michigan, Ann Arbor

Christos Faloutsos
Carnegie Mellon University

SYNTHESIS LECTURES ON DATA MINING AND KNOWLEDGE DISCOVERY #14

ABSTRACT

Graphs naturally represent information ranging from links between web pages, to communication in email networks, to connections between neurons in our brains. These graphs often span *billions* of nodes and interactions between them. Within this deluge of interconnected data, how can we find the most important structures and summarize them? How can we efficiently visualize them? How can we detect anomalies that indicate critical events, such as an attack on a computer system, disease formation in the human brain, or the fall of a company?

This book presents scalable, principled discovery algorithms that combine globality with locality to make sense of one or more graphs. In addition to fast *algorithmic methodologies*, we also contribute *graph-theoretical ideas and models*, and real-world *applications* in two main areas.

- **Individual Graph Mining**: We show how to interpretably *summarize* a single graph by identifying its important graph structures. We complement summarization with *inference*, which leverages information about few entities (obtained via summarization or other methods) and the network structure to efficiently and effectively learn information about the unknown entities.

- **Collective Graph Mining**: We extend the idea of individual-graph *summarization* to time-evolving graphs, and show how to scalably discover temporal patterns. Apart from summarization, we claim that *graph similarity* is often the underlying problem in a host of applications where multiple graphs occur (e.g., temporal anomaly detection, discovery of behavioral patterns), and we present principled, scalable algorithms for aligning networks and measuring their similarity.

The methods that we present in this book leverage techniques from diverse areas, such as matrix algebra, graph theory, optimization, information theory, machine learning, finance, and social science, to solve real-world problems. We present applications of our exploration algorithms to massive datasets, including a Web graph of 6.6 billion edges, a Twitter graph of 1.8 billion edges, brain graphs with up to 90 million edges, collaboration, peer-to-peer networks, browser logs, all spanning millions of users and interactions.

KEYWORDS

data mining, graph mining and exploration, graph similarity, graph matching, network alignment, graph summarization, pattern mining, outlier detection, anomaly detection, scalability, fast algorithms, models, visualization, social networks, brain graphs, connectomes

Contents

Acknowledgments xi

1 Introduction 1

1.1   Overview 1

1.2   Organization of This Book 2

    1.2.1   Part I: Individual Graph Mining 2

    1.2.2   Part II: Collective Graph Mining 3

    1.2.3   Code and Supporting Materials on the Web 5

1.3   Preliminaries 5

    1.3.1   Graph Definitions 5

    1.3.2   Graph-theoretic Data Structures 8

    1.3.3   Linear Algebra Concepts 9

    1.3.4   Select Graph Properties 11

1.4   Common Symbols 12

PART I   Individual Graph Mining **15**

2 Summarization of Static Graphs 17

2.1   Overview and Motivation 18

2.2   Problem Formulation 19

    2.2.1   MDL for Graph Summarization 21

    2.2.2   Encoding the Model 23

    2.2.3   Encoding the Errors 25

2.3   VoG: Vocabulary-based Summarization of Graphs 25

    2.3.1   Subgraph Generation 26

    2.3.2   Subgraph Labeling 26

    2.3.3   Summary Assembly 28

    2.3.4   Toy Example 29

    2.3.5   Time Complexity 29

2.4   Empirical Results 30

      2.4.1   Quantitative Analysis 31

      2.4.2   Qualitative Analysis 35

      2.4.3   Scalability 43

  2.5   Discussion 44

  2.6   Related Work 46

3   **Inference in a Graph** **49**

  3.1   Guilt-by-association Techniques 50

      3.1.1   Random Walk with Restarts (RWR) 50

      3.1.2   Semi-supervised Learning (SSL) 51

      3.1.3   Belief Propagation (BP) 51

      3.1.4   Summary 53

  3.2   FaBP: Fast Belief Propagation 53

      3.2.1   Derivation 58

      3.2.2   Analysis of Convergence 63

      3.2.3   Algorithm 64

  3.3   Extension to Multiple Classes 65

  3.4   Empirical Results 68

      3.4.1   Accuracy 69

      3.4.2   Convergence 69

      3.4.3   Robustness 70

      3.4.4   Scalability 70

PART II   Collective Graph Mining **73**

4   **Summarization of Dynamic Graphs** **75**

  4.1   Problem Formulation 77

      4.1.1   MDL for Dynamic Graph Summarization 79

      4.1.2   Encoding the Model 80

      4.1.3   Encoding the Errors 81

  4.2   TimeCrunch: Vocabulary-based Summarization of Dynamic Graphs 83

      4.2.1   Generating Candidate Static Structures 83

      4.2.2   Labeling Candidate Static Structures 84

      4.2.3   Stitching Candidate Temporal Structures 84

      4.2.4   Composing the Summary 86

  4.3   Empirical Results 87
       4.3.1   Quantitative Analysis 88
       4.3.2   Qualitative Analysis 90
       4.3.3   Scalability 92
     4.4   Related Work 93

5   Graph Similarity **97**
     5.1   Intuition 97
       5.1.1   Overview 99
       5.1.2   Measuring Node Affinities 99
       5.1.3   Leveraging Belief Propagation 100
       5.1.4   Desired Properties for Similarity Measures 101
     5.2   DELTACON: "δ" Connectivity Change Detection 102
       5.2.1   Algorithm Description 102
       5.2.2   Faster Computation 103
       5.2.3   Desired Properties 106
     5.3   DELTACON-ATTR: Adding Node and Edge Attribution 112
       5.3.1   Algorithm Description 113
       5.3.2   Scalability Analysis 115
     5.4   Empirical Results 115
       5.4.1   Intuitiveness of DELTACON 115
       5.4.2   Intuitiveness of DELTACON-ATTR 123
       5.4.3   Scalability 130
       5.4.4   Robustness 131
     5.5   Applications 132
       5.5.1   Enron 132
       5.5.2   Brain Connectivity Graph Clustering 134
       5.5.3   Recovery of Connectome Correspondences 135
     5.6   Related Work 138

6   Graph Alignment **143**
     6.1   Problem Formulation 144
     6.2   BIG-ALIGN: Bipartite Graph Alignment 146
       6.2.1   Mathematical Formulation 146
       6.2.2   Problem-specific Optimizations 149
       6.2.3   Algorithm Description 154
     6.3   UNI-ALIGN: Extension to Unipartite Graph Alignment 154

6.4   Empirical Results 157

     6.4.1   Accuracy and Runtime of BIG-ALIGN 157

     6.4.2   Accuracy and Runtime of UNI-ALIGN 161

6.5   Discussion 163

6.6   Related Work 163

7   **Conclusions and Further Research Problems** **167**

Bibliography **171**

Authors' Biographies **193**

Acknowledgments

This research was sponsored by the National Science Foundation under grant numbers IIS-1151017415, IIS-1217559, and IIS-1408924, the Department of Energy/National Nuclear Security Administration under grant number DE-AC52-07NA27344, the Defense Advanced Research Projects Agency under grant number W911NF-11-C-0088, the Air Force Research Laboratory under grant number F8750-11-C-0115, and the US Army Research Lab under grant number W911NF-09-2-0053. The views and conclusions contained in this document are those of the author and should not be interpreted as representing the official policies, either expressed or implied, of any sponsoring institution, the U.S. government or any other entity.

The authors would also like to thank all their collaborators for contributing to the works presented in this book: Polo Chau, Wolfgang Gatterbauer, Brian Gallagher, Stephan Guennemann, U Kang, Tai-You Ke, David Lubensky, Hsing-Kuo (Kenneth) Pao, Neil Shah, Hanghang Tong, Jilles Vreeken, and Joshua Vogelstein. We especially thank Neil Shah for his major contributions to Chapter 4: Summarization of Dynamic Graphs.

We also thank the reviewers of this book—Austin Benson, George Karypis, and Philip Yu—for their thoughtful feedback and suggestions which helped improve this book. Last but not least, we thank the informal reviewer, Haoming Shen, for carefully reading the manuscript, pointing out inconsistencies, and helping edit it.

Danai Koutra and Christos Faloutsos
October 2017

CHAPTER 1

Introduction

1.1 OVERVIEW

Graphs naturally represent information ranging from links between web pages, to users' movie preferences, to friendships and communications in social networks, to co-editor relationships in collaboration networks (Figure 1.1), to connections between voxels in our brains. Informally, a graph is a mathematical model for pairwise relations between objects. The objects are often referred to as nodes and the relations between them are represented by links or edges, which define influence and dependencies between the objects.

Figure 1.1: Wiki graph for the "Liancourt–Rocks" article, plotted using the "spring embedded" layout [101].

Real-world graphs often span hundreds of millions or even billions of nodes and interactions between them. Within this deluge of interconnected data, how can we extract *useful knowledge* in a scalable way and without flooding ourselves with unnecessary information? How can we find the most important structures and effectively summarize the graphs? How can we efficiently visualize them? How can we start from little prior information (e.g., few vandals and good contributors in Wikipedia) and broaden our knowledge to all the entities using network effects? How can we make sense of and explore multiple phenomena—represented as graphs—at the same time? How can we detect anomalies that indicate critical events, such as a cyber-attack or disease formation in the human brain? How can we summarize temporal graph patterns, such as the appearance and disappearance of an online community?

This book focuses on fast and principled methods for exploratory analysis of one or more networks in order to gain insights into the above-mentioned problems. The main directions of our work are: (a) summarization, which provides a compact and interpretable representation of one or more graphs, and (b) similarity, which enables the discovery of clusters of nodes or graphs with related properties. We provide theoretical underpinnings and scalable algorithmic approaches that exploit the sparsity in the data, and we show how to use them in large-scale, real-world applications, including anomaly detection in static and dynamic graphs (e.g., email communications or computer network monitoring), re-identification across networks, cross-network analytics, clustering, classification, and visualization.

1.2   ORGANIZATION OF THIS BOOK

This book is organized into two main parts: (i) individual graph mining and (ii) collective graph mining. We summarize the main problems of each part in the form of questions in Table 1.1. All these topics have broad impact on a variety of applications: anomaly detection in static and dynamic graphs, clustering and classification, cross-network analytics, re-identification across networks, and visualization in various types of networks, including social networks and brain graphs.

Table 1.1: Book organization

Part	Research Problem	Chapter
I: Individual Graph Mining	**Graph Summarization:** How can we succinctly describe a large-scale graph?	2
	Graph Mining Inference: What can we learn about all the nodes given prior information for a subset of them?	3
II: Collective Graph Mining	**Temporal Summarization:** How can we succinctly describe a set of large, temporal graphs?	4
	Graph Similarity: What is the similarity between two graphs? Which nodes and edges are responsible for their difference?	5
	Graph Alignment: How can we efficiently align two bipartite or unipartite graphs?	6

1.2.1   PART I: INDIVIDUAL GRAPH MINING

At a macroscopic level, how can we extract easy-to-understand building blocks from a massive graph and make sense of its underlying phenomena? At a microscopic level, after obtaining some knowledge about the graph structures, how can we further explore the nodes and find important node patterns (regular or anomalous)? The first part of the book introduces scalable ways for summarizing large-scale information by leveraging global and local graph properties. Summarization of massive data enables its efficient visualization, guides focus on its important aspects, and thus is key for understanding the data.

- Chapter 2: Summarization of Static Graphs—*"How can we succinctly describe a large-scale graph?"* A direct way of making sense of a graph is to model it at a macroscopic level and summarize it. Our method, VoG (Vocabulary-based summarization of Graphs) [123, 124], aims at succinctly describing a million-node graph with just a few, possibly-overlapping structures, which can be easily understood. We formalize the graph summarization problem as an information-theoretic optimization problem, where the goal is to find the hidden local structures that collectively minimize the global description length of the graph. In addition to leveraging the Minimum Description Length principle to find the best graph summary, another core idea is the use of a predefined vocabulary of structures that are semantically meaningful and ubiquitous in real networks: cliques and near-cliques, stars, chains, and (near-) bipartite cores.

- Chapter 3: Inference in a Graph (or Node similarity as Further Exploration)—*What can we learn about all the nodes given prior information for a subset of them?* After gaining knowledge about the important graph structures and their underlying behaviors through graph summarization (e.g., by using VoG), how can we extend our knowledge and find similar nodes within a graph, at a microscopic level? For example, suppose that we know a class-label (say, the type of contributor in Wikipedia, such as vandals/admins) for some of the nodes in the graph. Can we infer who else is a vandal in the network? This is the problem that we address in Chapter 3, in the case of two classes [125] and, its generalization to multiple [78] classes. The semi-supervised setting, where *some* prior information is available, appears in numerous domains, including law enforcement, fraud detection, and cyber security. Among the most successful methods that attempt to solve the problem are the ones that perform inference by exploiting the global network structure and local homophily effects ("birds of a feather flock together"). Starting from belief propagation, a powerful technique that handles both homophily and heterophily in networks, we have mathematically derived an accurate and faster (2×) linear approximation, FABP (Fast Belief Propagation), with convergence guarantees [125]. The derived formula revealed the equivalence of FABP to random walks with restarts and semi-supervised learning, and led to the *unification* of the three guilt-by-association methods.

1.2.2   PART II: COLLECTIVE GRAPH MINING

In many applications, it is necessary or at least beneficial to explore multiple graphs collectively. These graphs can be temporal instances on the same set of objects (dynamic or time-evolving graphs), or disparate networks coming from different sources. At a macroscopic level, how can we extract easy-to-understand building blocks from a *series of massive graphs* and *summarize* the dynamics of their underlying phenomena (e.g., communication patterns in a large phone-call network)? How can we find anomalies in a time-evolving corporate-email correspondence network and predict the fall of a company? Are there differences in the brain wiring of more creative and less creative people? How do different types of communication (e.g., messages vs. wall posts in Facebook) and their corresponding behavioral patterns compare? The second part of the book introduces scalable ways: (a) for summarizing large-scale temporal information by extending our ideas on single-graph summarization (Chapter 2), and (b) for comparing and aligning graphs, which are often the underlying problems in applications with multiple graphs.

- Chapter 4: Summarization of Dynamic Graphs—*How can we succinctly describe a set of large-scale, dynamic graphs?* Just like in the case of a single graph, a natural way of making sense of a series of graphs is to model them at a macroscopic level and summarize them. Our vocabulary-based method, TIMECRUNCH [189], succinctly describes a large, *time*-evolving graph with just a few phrases. Even visualizing a *single* large graph fails due to memory requirements or results in a clutter of nodes and edges without any useful information. Making sense of a large, time-evolving graph introduces even more challenges, so detecting simple temporal structures is crucial for visualization and understanding. Extending our work on single graph summarization presented in Chapter 2, we formalize the *temporal* graph summarization problem as an information-theoretic optimization problem, where the goal is to identify the temporal behaviors of local static structures that collectively minimize the global description length of the dynamic graph. We formulate a lexicon that describes various types of temporal behavior (e.g., flickering, periodic, one-shot) exhibited by the structures that we introduced for summarization of static graphs in Chapter 2 (e.g., stars, cliques, bipartite cores, chains).

- Chapter 5: Graph Similarity—*What is the similarity between two graphs? Which nodes and edges are responsible for their difference?* Graph similarity, i.e., the problem of assessing the similarity between two node-aligned graphs, has numerous high-impact applications, such as real-time anomaly detection in e-commerce/computer networks, which can prevent damage of millions of dollars. Although it is a long-studied problem, most methods do not give intuitive results and ignore the 'inherent' importance of the graph edges—e.g., an edge connecting two tightly-connected components is often assumed as important as an edge connecting two nodes in a clique. Our work [126, 128] redefines the space with new desired properties for graph similarity measures, and addresses scalability challenges. Based on the new requirements, we devised a massive-graph similarity algorithm, DELTACON (which stands for "δ connectivity" changes), which measures the differences in the k-step away neighborhoods in a principled way that uses a variant of Belief Propagation [125] introduced in Chapter 3 for inference in a single graph. DELTACON takes into account both local and global dissimilarities, but in a weighted manner: local dissimilarities (smaller k) are weighed higher than global ones (bigger k). It also detects the nodes and edges that are mainly responsible for the difference between the input graphs [126].

- Chapter 6: Graph Alignment—*How can we efficiently node-align two bipartite or unipartite graphs?* The graph similarity work introduced in Chapter 5 assumes that the correspondence of nodes across graphs is known, but this does not always hold true. Social network analysis, bioinformatics, and pattern recognition are just a few domains with applications that aim at finding node correspondence. In Chapter 6 we handle exactly this problem. Research has mostly focused on the alignment of *unipartite* networks. We focused on *bipartite* graphs, and formulated the alignment of such graphs as an optimization problem with practical constraints (e.g., sparsity, 1-to-many mapping) and developed a fast algorithm, BiG-ALIGN [127] (short for "Bipartite Graph Alignment"), to solve it. The key in solving the problem of global alignment efficiently is a series of optimizations that we devised, including the aggregation of nodes with local structural equivalence to supernodes. This leads to huge space and time savings: with careful handling, a node correspondence submatrix of several *millions* of small entries can be reduced to just *one*. Based on our formulation for bipartite graphs, we also introduced, UNI-ALIGN (short for "Unipartite Graph Alignment"), an alternative way of effectively and efficiently aligning unipartite graphs.

1.2.3   CODE AND SUPPORTING MATERIALS ON THE WEB

Research code and additional supplementary materials (e.g., slides) for the methods that we present in this book can be found at:

```
https://github.com/GemsLab/MCbook_Individual-Collective_GraphMining.git
```

1.3   PRELIMINARIES

In this section we introduce the main notions and definitions in graph theory and mining that are useful for understanding the methods and algorithms described in this book.

1.3.1   GRAPH DEFINITIONS

We start with the definition of a graph, followed by the different types of graphs (e.g., bipartite, directed, weighted), and the special cases of graphs or motifs (e.g., star, clique).

Graph   A representation of a set of objects connected by links (Figure 1.2). Mathematically, it is an ordered pair $G = (\mathcal{V}, \mathcal{E})$, where $\mathcal{V}$ is the set of objects (called nodes or vertices) and $\mathcal{E}$ is the set of links between some of the objects (also called edges).

Nodes or Vertices   A finite set $\mathcal{V}$ of objects in a graph. For example, in a social network, the nodes can be people, while in a brain network they correspond to voxels or cortical regions. The total number of nodes in a graph is often denoted as $|\mathcal{V}|$ or n.

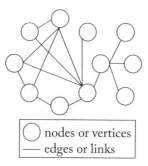

nodes or vertices
edges or links

Figure 1.2: An undirected, unweighted graph.

Edges or Links   A finite set $\mathcal{E}$ of connections between objects in a graph. The edges represent relationships between the objects—e.g., friendship between people in social networks, water flow between voxels in our brains. The total number of edges in a graph is often denoted as $|\mathcal{E}|$ or m.

Neighbors   Two vertices v and u connected by an edge are called neighbors. Vertex u is called the neighbor or adjacent vertex of v. In a graph G, the **neighborhood** of a vertex v is the induced subgraph of G consisting of all vertices adjacent to v and all edges connecting two such vertices.

Bipartite Graph   A graph that does not contain any odd-length cycles. Alternatively, a bipartite graph is a graph whose vertices can be divided into two disjoint sets $\mathcal{U}$ and $\mathcal{V}$ such that every edge connects a vertex in $\mathcal{U}$ to one in $\mathcal{V}$ (Figure 1.3). A graph whose vertex sets cannot be divided into disjoint sets with that property is called **unipartite**. A tree is a special case of bipartite graph.

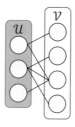

Figure 1.3: A bipartite graph with three nodes in set $\mathcal{U}$ and 4 nodes in set $\mathcal{V}$.

Directed Graph   A graph whose edges have a direction associated with them (Figure 1.4). A directed edge is represented by an ordered pair of vertices (u, v), and is illustrated by an arrow starting from u and ending at v. A directed graph is also called digraph or directed network. The directionality captures non-reciprocal relationships. Examples of directed networks are the who-follows-whom Twitter network (an arrow starts from the follower and ends at the followee) or a phonecall network (with arrows from the caller to the callee). A graph whose edges are unordered pairs of vertices is called **undirected**.

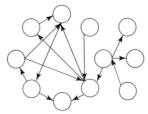

Figure 1.4: A directed graph.

Weighted Graph   A graph whose edges have a weight associated with them (Figure 1.5). If the weights of all edges are equal to 1, then the graph is called **unweighted**. The weights can be positive or negative, integers or decimal. For example, in a phonecall network the weights may represent the number of calls between two people.

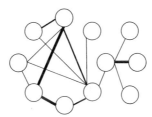

Figure 1.5: A weighted graph. The edge width is proportional to the edge weight.

Labeled Graph   A graph whose nodes or edges have a label associated with them (Figure 1.6). An example of a vertex-labeled graph is a social network where the names of the people are known.

Egonet   The egonet of node v is the induced subgraph of G which contains v, its neighbors, and all the edges between them. Alternatively, it is the 1-step neighborhood of node v.

Simple Graph   An undirected, unweighted graph containing no loops (edges from a vertex to itself) or multiple edges.

Clique or Complete Graph   A clique is a set of vertices in a graph, such that every two distinct vertices are adjacent. The induced subgraph of a clique is called a complete graph.

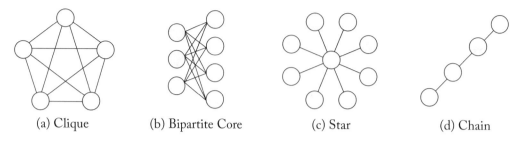

| (a) Clique | (b) Bipartite Core | (c) Star | (d) Chain |

Figure 1.6: Special cases of graphs.

Bipartite Core or Complete Bipartite Graph   A special case of bipartite graph where every vertex of the first set ($\mathcal{U}$) is connected to every vertex of the second set $\mathcal{V}$. It is denoted as $K_{s,t}$, where s and t are the number of vertices in $\mathcal{U}$ and $\mathcal{V}$, respectively.

Star   A complete bipartite graph $K_{1,t}$, for any t. The vertex in set $\mathcal{U}$ is called the central node or hub, while the vertices in $\mathcal{V}$ are often called peripheral nodes or spokes.

Chain   A graph that can be drawn so that all of its vertices and edges lie on a single straight line.

Triangle   A 3-node complete graph.

1.3.2   GRAPH-THEORETIC DATA STRUCTURES

The data structure used for a graph depends on its properties (e.g., sparse, dense, small, large) and the algorithm applied to it. Matrices have big memory requirements, and thus are preferred for small, dense graphs. On the other hand, lists are better for large, sparse graphs, such as social, collaboration, and other real-world networks.

Adjacency matrix   The adjacency matrix of a graph G is an $n \times n$ matrix $\mathbf{A}$ (or else $\mathbf{A} \in \mathbb{R}^{n \times n}$), whose element a_{ij} is non-zero if vertex i is connected to vertex j, and 0 otherwise. In other words, it represents which vertices of the graph are adjacent to which other vertices. For a **graph without loops**, the diagonal elements a_{ii} are 0. The adjacency matrix of an **undirected graph** is symmetric. The elements of the adjacency matrix of a **weighted graph** are equal to the weights of the corresponding edges (Figure 1.7).

The $n \times n$ adjacency matrix of a bipartite graph with two sets of nodes $\mathcal{V}_1$ and $\mathcal{V}_2$ is of the form $\mathbf{A} = \left(\begin{array}{c|c} \mathbf{0} & \mathbf{B} \\ \hline \mathbf{B}^T & \mathbf{0} \end{array} \right)$, where $\mathbf{B}$ is a $n_1 \times n_2$ matrix with the connections between the n_1 nodes in set $\mathcal{V}_1$ and the n_2 nodes in $\mathcal{V}_2$, and $\mathbf{B}^T$ is the transpose matrix of $\mathbf{B}$.

Incidence list   An array of pairs of adjacent vertices. This is a common representation for sparse, real-world networks, because of its efficiency in terms of memory and computation.

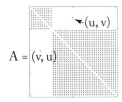

Figure 1.7: Adjacency matrix of a simple, unweighted, and undirected graph.

Sparse matrix A matrix in which most of the elements are zero. A matrix where most of the entries are non-zero is called **dense**. Most of the real-world networks (e.g., social, collaboration, and phonecall networks) are sparse.

Degree matrix An $n \times n$ diagonal matrix $\mathbf{D}$ that contains the degree of each node and corresponds to the $n \times n$ adjacency matrix $\mathbf{A}$. Its ith element, $d_{ii} = \sum_{j=1}^{n} a_{ij}$, represents the degree of node i, $d(i)$. The adjacency and degree matrices of a simple, unweighted and undirected chain graph is given in Figure 1.8.

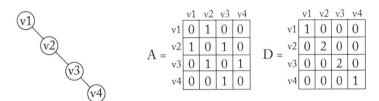

Figure 1.8: A chain of four nodes, and its adjacency and degree matrices.

Laplacian matrix An $n \times n$ matrix defined as $\mathbf{L} = \mathbf{D} - \mathbf{A}$, where $\mathbf{A}$ is the adjacency matrix of the graph, and $\mathbf{D}$ is the degree matrix. The Laplacian matrix arises in the analysis of random walks, electrical networks on graphs, spectral clustering, and other graph applications.

1.3.3 LINEAR ALGEBRA CONCEPTS

Since graphs are represented by their corresponding matrices, graph problems involve linear algebra operations. Here we review some basic linear algebra concepts, which are used in the methods that we describe in the following chapters. Below we give definitions for a square $n \times n$ matrix $\mathbf{A}$ with elements a_{ij}, unless mentioned otherwise.

Transpose The transpose of a matrix $\mathbf{A} \in \mathbb{R}^{n \times p}$ is a matrix $\mathbf{C}$ with elements $c_{ij} = a_{ji}$, for $i = 1, \ldots, n$; $j = 1, \ldots, p$. The transpose is denoted by $\mathbf{A}^T$. Note that the transpose is defined for any matrix, not only square matrices.

Diagonal Matrix   A diagonal matrix has all its non-zero entries on the main diagonal. Although it usually refers to square matrices, it is defined for rectangular matrices as well. The degree matrix $\mathbf{D}$ of a graph is an example of a diagonal matrix (Figure 1.8).

Identity Matrix   The identity matrix $\mathbf{I}$ is a square diagonal matrix with ones on the main diagonal and zeros elsewhere.

Orthogonal Matrix   A square matrix $\mathbf{A} \in \mathbb{R}^{n \times n}$ is called orthogonal if its columns and rows are orthogonal unit vectors, or $\mathbf{A}\mathbf{A}^T = \mathbf{I}$.

Inverse   The inverse of a square matrix $\mathbf{A} \in \mathbb{R}^{n \times n}$, if it exists, is a matrix $\mathbf{C}$ such that $\mathbf{A}\mathbf{C} = \mathbf{C}\mathbf{A} = \mathbf{I}$. The inverse is denoted by $\mathbf{A}^{-1}$.

Eigenvalue Decomposition or Eigendecomposition   A scalar number λ is an eigenvalue of a square matrix $\mathbf{A} \in \mathbb{R}^{n \times n}$ if there is a non-zero vector $\mathbf{x}$ such that $\mathbf{A}\mathbf{x} = \lambda\mathbf{x}$. The vector $\mathbf{x}$ is the eigenvector corresponding to the eigenvalue λ. A matrix has at most n distinct eigenvalues. The eigenvalue with the maximum magnitude is called spectral radius, and is denoted by $\rho(\mathbf{A}) = max_i |\lambda_i|$.

Trace   The trace of a square matrix $\mathbf{A} \in \mathbb{R}^{n \times n}$ is equal to the sum of its diagonal elements, i.e., $Tr(\mathbf{A}) = \sum_{i=1}^{n} a_{ii}$. The trace of a square matrix is equal to the sum of its eigenvalues.

Rank   The rank of a matrix is equal to the number of its linearly independent columns. A matrix $\mathbf{A} \in \mathbb{R}^{n \times p}$ is full rank when its rank is equal to the smallest of n and p.

Matrix A^k   Let a square matrix $\mathbf{A} \in \mathbb{R}^{n \times n}$ represent a graph G. The kth power of $\mathbf{A}$ has elements $(A^k)_{ij}$, which represent the length-k paths connecting the nodes i and j. In this case, a path can repeat both nodes and edges (also referred to as walks).

Vector Norms   Vector norms are functions that operate on vectors $\mathbf{x} \in \mathbb{R}^n$ and assign to them a positive length. The most commonly used vector norms are special cases of the p-norm: $||\mathbf{x}||_p = \left(\sum_{i=1}^{n} |x_i|^p \right)^{\frac{1}{p}}$. The cases of $p = 1, 2$ and ∞ result in the most widely used norms: the Taxicab or Manhattan or l_1 norm; the Euclidean norm or l_2 norm; and the infinity or maximum norm $||\mathbf{x}||_\infty = max_i |x_i|$. The zero "norm", l_0, represents the number of non-zeroes in a vector. This does not constitute a proper norm, but it is often used in graph-related problems (and relaxed to mathematically tractable formulations).

Matrix Norms   These are extensions of the vector norms to matrices. One of the most popular matrix norms is the Frobenius norm, which is defined as follows for matrix $\mathbf{A} \in \mathbb{R}^{n \times p}$: $||\mathbf{A}||_F = \sqrt{Tr(\mathbf{A}^T \mathbf{A})} = \sqrt{\sum_{i=1}^{n} \sum_{j=1}^{p} |a_{ij}|^2}$.

Singular Value Decomposition (SVD)   The singular value decomposition is a type of matrix factorization for any matrix $\mathbf{A} \in \mathbb{R}^{n \times p}$. It is a generalization of the eigenvalue decomposition, which applies only to square matrices (specific instances). The SVD of $\mathbf{A}$ is given in the form $\mathbf{A} = \mathbf{U}\mathbf{\Sigma}\mathbf{V}^T$, where $\mathbf{U}$ is an $n \times n$ orthogonal matrix with the left singular vectors, $\mathbf{V}$ is a $p \times p$ orthogonal matrix with the right singular vectors, and $\mathbf{\Sigma}$ is a $n \times p$ diagonal matrix with non-negative real numbers on the diagonal. The singular values of matrix $\mathbf{A}$ are equal to the square root of the non-zero eigenvalues of the square matrix $\mathbf{A}^T\mathbf{A}$ (or $\mathbf{A}\mathbf{A}^T$).

SVD is often used to obtain a rank-k approximation of a matrix, which can speed up graph algorithms significantly. A rank-k approximation of a matrix is given as $\hat{\mathbf{A}} = \mathbf{U}\hat{\mathbf{\Sigma}}\mathbf{V}^T$, where $\hat{\mathbf{\Sigma}}$ has only the top-k singular values on the diagonal and the others are set to 0.

The time complexity of exact SVD is $O(\min\{n^2 p, np^2\})$, or $O(n^3)$ for a square matrix. Randomized methods allow to compute (good approximations of) the top-k singular values of a matrix significantly faster.

Power Method   The power method or power iteration is used to find the eigenvalues of a matrix efficiently, while avoiding expensive matrix decompositions. Especially in the case of real-world graphs, their corresponding matrices are sparse (i.e., they have few non-zero elements). The key idea of the power method is to rely on fast sparse-matrix and vector multiplications, which are computed in iterations. For instance, for a random (non-zero) vector $\mathbf{x}$ and a matrix $\mathbf{A}$, the power method would iteratively compute $\mathbf{A}\mathbf{x}$, $\mathbf{A}^2\mathbf{x}$, $\mathbf{A}^3\mathbf{x}$, etc. In this book, we use the idea of the power method to efficiently invert a matrix in Chapter 3, so we do not give the details of computing the eigenvalues of a matrix.

1.3.4   SELECT GRAPH PROPERTIES

A graph and its nodes are characterized by various properties or invariants or features. In this section we review some important properties that are used throughout the book.

Degree   The degree of a vertex v (denoted as $d(v)$) in a graph G is the number of edges incident to the vertex, i.e., the number of its neighbors. For an unweighted graph with a corresponding $n \times n$ binary matrix $\mathbf{A}$, the degree of each vertex is $d(i) = \sum_{j=1}^{n} a_{ij}$. In a directed graph, the **in-degree** of a vertex is the number of incoming edges, and its **out-degree** is the number of outgoing edges. Often, the degrees of the vertices in a graph are represented compactly as a diagonal matrix $\mathbf{D}$, where $d_{ii} = d(i)$. The **degree distribution** is the probability distribution of the node degrees in graph G.

PageRank   The PageRank of a node v is a score that captures its importance relevant to the other nodes. The score depends only on the graph structure. PageRank is the algorithm used by Google Search to rank web pages in the search results [35].

Geodesic Distance   The geodesic distance between two vertices v and u is the length of the shortest path between them. It is also called **hop count or distance**.

Node Eccentricity or Radius   The eccentricity or radius of node v is the greatest geodesic distance between v and any other vertex in the graph. Intuitively, eccentricity captures how far a node is from the furthest away vertex in the graph.

Graph Diameter   The diameter of a graph is the maximum eccentricity of any node in the graph. The smallest eccentricity over all the vertices in the graph is called graph **radius**. In Figure 1.6d, the diameter of the chain is 3, and its radius is 2.

Connected Component   In an undirected graph, a connected component is a subgraph in which any vertex is reachable from any other vertex (i.e., any two vertices are connected to each other by paths), and which is connected to no additional vertices in the graph. A vertex without neighbors is itself a connected component. The number of connected components in a graph is equal to the multiplicity of 0 as an eigenvalue of its Laplacian matrix **L**. Intuitively, in co-authorship networks, a connected component corresponds to researchers who publish together, while different components may represent groups of researchers in different areas who have never published papers together.

Participating Triangles   The participating triangles of a node are defined as the number of distinct triangles in which it participates. Triangles have been used for spam detection, link prediction, and recommendation in social and collaboration networks, and other real-world applications.

Eigenvectors and Eigenvalues   The eigenvalues (eigenvectors) of a graph G are defined to be the eigenvalues (eigenvectors) of its corresponding adjacency matrix, **A**. The eigenvalues characterize the graph's topology and connectedness (e.g., bipartite, complete graph), and are often used to count various subgraph structures, such as spanning trees. The eigenvalues of undirected graphs (with symmetric adjacency matrices) are real. The (first) principal eigenvector captures the centrality of the graph's vertices—this is related to Google's PageRank algorithm. The second smallest eigenvector is used for graph partitioning via spectral clustering.

1.4   COMMON SYMBOLS

We give the most common symbols that we use throughout this book and their short descriptions in Table 1.2. Additional symbols necessary to explain individual methods and algorithms are provided in the corresponding chapters.

Table 1.2: Common symbols and definitions. Bold uppercase letters for matrices; bold lowercase letters for vectors; plain font for scalars.

Symbol	Description
$G(\mathcal{V}, \mathcal{E})$	Graph
$\mathcal{V}, n = |\mathcal{V}|$	Node-set and number of notes of G, respectively
$\mathcal{E}, m = |\mathcal{E}|$	Edge-set and number of edges of G, respectively
$G_x(\mathcal{V}_x, \mathcal{E}_x)$	x^{th} graph in collective graph mining
$\mathcal{V}_x, n_x = |\mathcal{V}_x|$	Node-set and number of nodes of G_x, respectively
$\mathcal{E}_x, m_x = |\mathcal{E}_x|$	Edge-set and number of edges of G_x, respectively
$\mathbf{A}$	Adjacency matrix of graph G with elements $a_{ij} \in \mathbb{R}$
$\mathbf{A}_x$	Adjacency matrix of graph G_x
$\mathbf{I}$	$n \times n$ identity matrix
$\mathbf{D}$	$n \times n$ diagonal degree matrix, $d_{ii} = \sum_j a_{ij}$ and $d_{ij} = 0$ for $i \neq j$
$\mathbf{L}$	$= \mathbf{D} - \mathbf{A}$ Laplacian matrix
$\mathbf{L}_{norm}$	$\mathbf{D}^{-1/2}\mathbf{L}\mathbf{D}^{-1/2}$ normalized Laplacian matrix

PART I

Individual Graph Mining

CHAPTER 2

Summarization of Static Graphs

One natural way to understand a graph and its underlying processes is to visualize and interact with it. However, for large datasets with several millions or billions of nodes and edges, such as the Facebook social network, even loading them using an appropriate visualization software requires significant amount of time. If the memory requirements are met, visualizing the graph is possible, but the result is a "hairball" without obvious patterns: often the number of nodes is larger than the number of pixels on a screen, while, at the same time, people have limited capacity for processing information. How can we summarize efficiently, and in simple terms, which parts of the graph stand out? What can we say about its structure? Its edge distribution will likely follow a power law [64], but apart from that, is it random? The focus of this chapter is finding short summaries for large graphs, in order to gain a better understanding of their characteristics.

Why not apply one of the many community detection, clustering or graph-cut algorithms that abound in the literature [41, 50, 109, 134, 176], and summarize the graph in terms of its communities? The answer is that these algorithms do not quite serve our goal. Typically, they detect numerous communities without explicit ordering, so a principled selection procedure of the most "important" subgraphs is still needed. In addition to that, these methods merely return the discovered communities, without characterizing them (e.g., clique, star), and thus, do not help the user to gain further insights into the properties of the graph.

We introduce VoG or *Vocabulary-based summarization of Graphs*, an efficient and effective method for summarizing and understanding large real-world graphs. In particular, we aim at understanding graphs *beyond* the so-called caveman networks[1] that only consist of well-defined, tightly-knit clusters, which are known as cliques and near-cliques in graph terms.

The first insight is to best *describe* the structures in a graph using an enriched set of "vocabulary" terms: cliques and near-cliques (which are typically considered by community detection methods), and also stars, chains, and (near) bipartite cores. The reasons we chose these "vocabulary" terms are: (a) (near-) cliques are included, and so our method works fine on caveman graphs, and (b) stars [139], chains [206], and bipartite cores [117, 176] appear very often, and

---

[1]A caveman graph arises by modifying a set of fully connected clusters (caves) by removing one edge from each cluster and using it to connect to a neighboring one such that the clusters form a single loop [220]. Intuitively, a caveman graph has a block-diagonal matrix, with a few edge additions and deletions.

have semantic meaning (e.g., factions, bots) in the tens of real networks we have seen in practice (e.g., the IMDB movie-actor graph, co-authorship networks, Netflix movie recommendations, US Patent dataset, phonecall networks).

The second insight is to *formalize* our goal using the Minimum Description Length (MDL) principle [183] as a lossless compression problem. That is, by MDL we define the best summary of a graph as the set of subgraphs that describes the graph most succinctly, i.e., compresses it best, and, thus, may help a human understand the main graph characteristics in a simple, non-redundant manner. A big advantage is that VoG is *parameter-free*, as at any stage MDL identifies the best choice: the one by which we save most bits.

Informally, we present a scalable solution to the following problem.

Problem 2.1   Graph Summarization - Informal   Given a graph, find a set of possibly overlapping subgraphs **to most succinctly describe** the given graph, i.e., explain as many of its edges in as simple terms as possible, in a **scalable** way, ideally linear on the number of edges.

2.1   OVERVIEW AND MOTIVATION

Before we present the problem formulation and the search algorithm, we first provide the high-level outline of VoG:

(a) We use MDL to formulate a quality function: a collection M of structures (e.g., a star here, cliques there, etc.) is as good as its description length $L(G, M)$. Hence, any subgraph or set of subgraphs has a quality score.

(b) We give an efficient algorithm for characterizing candidate subgraphs. In fact, we allow *any* subgraph discovery heuristic to be used for this, as we define our framework in general terms and use MDL to identify the structure *type* of the candidates.

(c) Given a candidate set C of promising subgraphs, we show how to mine informative summaries, removing redundancy by minimizing the compression cost.

VoG results in a list M of, possibly overlapping subgraphs, sorted in order of importance (compression gain). Together these subgraphs succinctly describe the main connectivity of the graph.

The motivation behind VoG is that the visualization of large graphs often results in a clutter of nodes and edges, and hinders interactive exploration and discoveries. On the other hand, a handful of simple, "important" structures can be visualized more easily, and may help the user understand the underlying characteristics of the graph. Next, we give an illustrating example of VoG, where the most 'important' vocabulary subgraphs that constitute a Wikipedia article's (graph) summary are semantically interesting.

Illustrating Example   In Figure 2.1 we give the results of VoG on a Wikipedia graph based on the article about Liancourt-Rocks; the nodes are editors, and editors share an edge if they edited the same part of the article. Figure 2.1a shows the graph using the spring-embedded model [101]. No clear pattern emerges, and thus a human would have hard time understanding this graph. Contrast that with the results of VoG. Figures 2.1b–2.1d depict the same graph, where we highlight the most important structures (i.e., structures that save the most bits) discovered by VoG. The discovered structures correspond to behavioral patterns.

- *Stars → admins (+ vandals)*: In Figure 2.1b, with red color, we show the centers of the most important "stars": further inspection shows that these centers typically correspond to administrators who revert vandalisms and make corrections.

- *Bipartite cores → edit wars*: Figures 2.1c and 2.1d give the two most important near-bipartite-cores. Manual inspection shows that these correspond to *edit wars*: two groups of editors reverting each others' changes. For clarity, we denote the members of one group by red nodes (left), and highlight the edges to the other group in pale yellow.

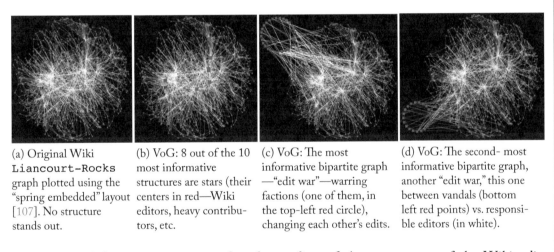

(a) Original Wiki Liancourt-Rocks graph plotted using the "spring embedded" layout [107]. No structure stands out.

(b) VoG: 8 out of the 10 most informative structures are stars (their centers in red—Wiki editors, heavy contributors, etc.

(c) VoG: The most informative bipartite graph —"edit war"—warring factions (one of them, in the top-left red circle), changing each other's edits.

(d) VoG: The second- most informative bipartite graph, another "edit war," this one between vandals (bottom left red points) vs. responsible editors (in white).

Figure 2.1: VoG: summarization and understanding of those structures of the Wikipedia Liancourt-Rocks graph that are most important from an information-theoretic point of view. Nodes stand for Wikipedia contributors and edges link users who edited the same part of the article.

2.2 PROBLEM FORMULATION

In this section we describe the first contribution, the MDL formulation of graph summarization. To enhance readability, we list the most frequently used symbols in Table 2.1.

Table 2.1: VoG: Description of the major symbols for static graph summarization

Symbol	Description
fc, nc	*Full* clique and *near* clique, respectively
fb, nb	Full bipartite core and *near* bipartite core, respectively
st	Star graph
ch	Chain graph
Ω	Vocabulary of structure types, e.g., $\Omega \subseteq \{fc, nc, fr, nr, fb, nb, ch, st\}$
$\mathcal{C}_x$	Set of all candidate structures of type $x \in \Omega$
$\mathcal{C}$	Set of all candidate structures, $\mathcal{C} = \cup_x \mathcal{C}_x$
M	A model for G, essentially a list of node sets with associated structure types
$s, t \in M$	Structures in M
area (s)	Edges of G (= cells of **A**) described by s
$|S|, |s|$	Cardinality of set S and number of nodes in s, respectively
$\|s\|, \|s\|'$	Number of existing, resp. non-existing edges within the area of **A** that s describes
M	Approximation of adjacency matrix **A** deduced by M
E	Error matrix, $\mathbf{E} = \mathbf{M} \oplus \mathbf{A}$
$\oplus$	Exclusive OR
$L(G, M)$	Number of bits to describe model M, and G using M
$L(M)$	Number of bits to describe model M
$L(s)$	Number of bits to describe structure s

In general, the Minimum Description Length principle (MDL) [184] is a practical version of Kolmogorov Complexity [138], which embraces the slogan *Induction by Compression*. For MDL, this can be roughly described as follows. Given a set of models $\mathcal{M}$, the best model $M \in \mathcal{M}$ minimizes

$$L(M) + L(\mathcal{D} \mid M) ,$$

where

- $L(M)$ is the length, in bits, of the description of M, and

- $L(\mathcal{D} \mid M)$ is the length, in bits, of the description of the data when encoded using the information in M.

This is called two-part or *crude* MDL, as opposed to *refined* MDL, where model and data are encoded together [88]. We use two-part MDL because we are specifically interested in the model: those graph connectivity structures that together best describe the graph. Further, refined MDL cannot be computed except for some special cases [89], despite having stronger theoretical foundations.

Without loss of generality, we here consider undirected graphs $G(\mathcal{V}, \mathcal{E})$ of $n = |\mathcal{V}|$ nodes, and $m = |\mathcal{E}|$ edges, with no self-loops. Our theory can be straightforwardly generalized to directed graphs—and similarly so for weighted graphs, has an expectation or is willing to make an assumption on the distribution of the edge weights.

To use MDL for graph summarization, we need to define what our models $\mathcal{M}$ are, how a model $M \in \mathcal{M}$ describes data, and how we encode this in bits. We do this next. It is important to note that to ensure fair comparison, MDL requires descriptions to be lossless, and, that in MDL we are only concerned with the optimal description *lengths*—not actual instantiated code words—and hence do not have to round up to the nearest integer.

2.2.1 MDL FOR GRAPH SUMMARIZATION

As models M, we consider ordered lists of graph structures. We write Ω for the set of graph structure *types* that are allowed in M, i.e., that we are allowed to describe (parts of) the input graph with. We will colloquially refer to Ω as our *vocabulary*. Although in principle any graph structure type can be a part of the vocabulary, we here choose the six most common structures in real-world graphs [117, 176, 206] that are well known and understood by the graph mining community: *full* and *near* cliques (*fc, nc*), *full* and *near* bipartite cores (*fb, nb*), stars (*st*), and chains (*ch*). Compactly, we have $\Omega = \{fc, nc, fb, nb, ch, st\}$. We will formally introduce these types after formalizing our goal.

Each structure $s \in M$ identifies a patch of the adjacency matrix $\mathbf{A}$ and describes how it is connected (Figure 2.2). We refer to this patch, or more formally the edges $(i, j) \in \mathbf{A}$ that structure s describes, as $area(s, M, \mathbf{A})$, where we omit M and $\mathbf{A}$ whenever clear from context.

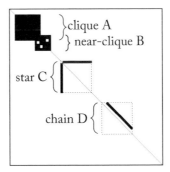

Figure 2.2: Illustration of our main idea on a toy adjacency matrix: VoG identifies *overlapping* sets of nodes, that form vocabulary subgraphs (cliques, stars, chains, etc). VoG allows for the soft clustering of nodes, as in clique A and near-clique B. Stars look like inverted L shapes (e.g., star C). Chains look like lines parallel to the main diagonal (e.g., chain D).

We allow overlap between structures[2]: nodes may be part of more than one structure. We allow, for example, cliques to overlap. Edges, however, are described on a first-come-first-serve basis: the first structure $s \in M$ to describe an edge (i, j) determines the value in $\mathbf{A}$. We do not impose constraints on the amount of overlap; MDL will decide for us whether adding a structure to the model is too costly with respect to the number of edges it helps to explain.

Let $\mathcal{C}_x$ be the set of all possible subgraphs of up to n nodes of type $x \in \Omega$, and $\mathcal{C}$ the union of all of those sets, $\mathcal{C} = \cup_x \mathcal{C}_x$. For example, $\mathcal{C}_{fc}$ is the set of all possible full cliques. Our model family $\mathcal{M}$ then consists of all possible permutations of all possible subsets of $\mathcal{C}$—recall that the models M are *ordered* lists of graph structures. By MDL, we are after the $M \in \mathcal{M}$ that best balances the complexity of encoding both $\mathbf{A}$ and M.

Our general approach for transmitting the adjacency matrix is as follows. First, we transmit the model M. Then, given M, we can build the approximation $\mathbf{M}$ of the adjacency matrix, as defined by the structures in M; we simply iteratively consider each structure $s \in M$, and fill out the connectivity of *area(s)* in $\mathbf{M}$ accordingly. As M is a summary, it is unlikely that $\mathbf{M} = \mathbf{A}$. Still, in order to fairly compare between models, MDL requires an encoding to be lossless. Hence, besides M, we also need to transmit the error matrix $\mathbf{E}$, which encodes the error with respect to $\mathbf{A}$. We obtain $\mathbf{E}$ by taking the exclusive OR between $\mathbf{M}$ and $\mathbf{A}$, i.e., $\mathbf{E} = \mathbf{M} \oplus \mathbf{A}$. Once the recipient knows M and $\mathbf{E}$, the full adjacency matrix $\mathbf{A}$ can be reconstructed without loss.

With this in mind, we have as our main score

$$L(G, M) = L(M) + L(\mathbf{E}),$$

where $L(M)$ and $L(\mathbf{E})$ are the numbers of bits that describe the structures, and the error matrix $\mathbf{E}$, respectively. We note that $L(\mathbf{E})$ maps to $L(\mathcal{D} \mid M)$, introduced in Section 2.2. That is, it corresponds to the length, in bits, of the description of the data when encoded, using the information in M. The formal definition of the problem we tackle in this work is defined as follows.

Problem 2.2 Minimum Graph Description Problem   Given a graph G with adjacency matrix A, and the graph structure vocabulary Ω, by the MDL principle we are after the smallest model M for which the total encoded length is minimal, that is

$$\min L(G, M) = \min\{L(M) + L(\mathbf{E})\},$$

where $\mathbf{E} = \mathbf{M} \oplus \mathbf{A}$ is the error matrix, and $\mathbf{M}$ is an approximation of $\mathbf{A}$ deduced by M.

Next, we formalize the encoding of the model and the error matrix.

---

[2]This is a common assumption in mixed-membership stochastic blockmodels [4, 107].

2.2.2   ENCODING THE MODEL

For the encoded length of a model $M \in \mathcal{M}$, we have

$$
L(M) \;=\; \underbrace{L_{\mathbb{N}}(|M|+1) + \log \binom{|M|+|\Omega|-1}{|\Omega|-1}}_{\text{\# of structures, in total, and per type}} + \underbrace{\sum_{s \in M} \left(-\log \Pr(x(s) \mid M) + L(s) \right)}_{\text{per structure, in order, type and details}} \quad.
$$

First, we transmit the total number of structures in M using $L_{\mathbb{N}}$, the MDL optimal encoding for integers greater than or equal to 1 [184]. Next, by an index over a weak number composition, we optimally encode the number of structures of each type $x \in \Omega$ in model M. Then, for each structure $s \in M$, we encode its type $x(s)$ with an optimal prefix code [52], and finally its structure.

     To compute the encoded length of a model, we need to define $L(s)$ per graph structure type in our vocabulary.

Cliques

To encode a *full clique*, a set of fully connected nodes as a *full clique*, we first encode the number of nodes, and then their IDs

$$
L(fc) = \underbrace{L_{\mathbb{N}}(|fc|)}_{\text{\# of nodes}} + \underbrace{\log \binom{n}{|fc|}}_{\text{node IDs}}.
$$

For the number of nodes we re-use $L_{\mathbb{N}}$, and we encode their IDs by an index over an ordered enumeration of all possible ways to select $|fc|$ nodes out of n. As M generalizes the graph, we do not require that fc is a full clique in G. If only few edges are missing, it may still be convenient to describe it as such. Every missing edge, however, adds to the cost of transmitting $\mathbf{E}$.

    Less dense or *near*-cliques can be as interesting as full-cliques. We encode these as follows:

$$
L(nc) \;=\; \underbrace{L_{\mathbb{N}}(|nc|)}_{\text{\# of nodes}} + \underbrace{\log \binom{n}{|nc|}}_{\text{node IDs}} + \underbrace{\log(|area(nc)|)}_{\text{\# of edges}} + \underbrace{||nc||l_1 + ||nc||'l_0}_{\text{edges}}.
$$

We first transmit the number and IDs of nodes as above, and then identify which edges are present and which are not, using optimal prefix codes. We write $||nc||$ and $||nc||'$ for respectively the number of present and missing edges in $area(nc)$. Then, $l_1 = -\log((||nc||/(||nc||+||nc||'))$, and analogue for l_0, are the lengths of the optimal prefix codes for present and missing edges, respectively. The intuition is that the more dense (sparse) a near-clique is, the cheaper encoding its edges will be. Note that this encoding is exact; no edges are added to $\mathbf{E}$.

Bipartite Cores

Bipartite cores are defined as non-empty, non-intersecting sets of nodes, A and B, for which there are edges only *between* the sets A and B, and not *within*.

The encoded length of a full bipartite core *fb* is

$$L(fb) = \underbrace{L_{\mathbb{N}}(|A|) + L_{\mathbb{N}}(|B|)}_{\text{cardinality of } A \text{ and } B, \text{ resp.}} + \underbrace{\log \binom{n}{|A|}}_{\text{node IDs in } A} + \underbrace{\log \binom{n - |A|}{|B|}}_{\text{node IDs in } B},$$

where we encode the size of A, B, and then the node IDs.

Analogue to cliques, we also consider near bipartite cores, *nb*, where the core is not (necessarily) fully connected. To encode a near bipartite core we have

$$L(nb) = \underbrace{L_{\mathbb{N}}(|A|) + L_{\mathbb{N}}(|B|)}_{\text{cardinality of } A \text{ and } B, \text{ resp.}} + \underbrace{\log \binom{n}{|A|}}_{\text{node IDs in } A} + \underbrace{\log \binom{n - |A|}{|B|}}_{\text{node IDs in } B}$$

$$+ \underbrace{\log(|area(nb)|)}_{\text{number of edges}} + \underbrace{||nb|| l_1 + ||nb||' l_0}_{\text{edges}}.$$

Stars

A star is specific case of the bipartite core that consists of a single node (hub) in A connected to a set B of at least 2 nodes (spokes). For $L(st)$ of a given star *st* we have

$$L(st) = \underbrace{L_{\mathbb{N}}(|st| - 1)}_{\text{number of spokes}} + \underbrace{\log n}_{\text{id of hub node}} + \underbrace{\log \binom{n - 1}{|st| - 1}}_{\text{ids of spoke nodes}},$$

where $|st| - 1$ is the number of spokes of the star. To identify the member nodes, we first identify the hub out of n nodes, and then the spokes from the remaining nodes.

Chains

A chain is a list of nodes such that every node has an edge to the next node, i.e., under the right permutation of nodes, **A** has only the super-diagonal elements (directly above the diagonal) non-zero. As such, for the encoded length $L(ch)$ for a chain *ch* we have

$$L(ch) = \underbrace{L_{\mathbb{N}}(|ch| - 1)}_{\text{\# of nodes in chain}} + \underbrace{\sum_{i=0}^{|ch|} \log(n - i)}_{\text{node IDs, in order of chain}},$$

where we first encode the number of nodes in the chain, and then their IDs in order. Note that $\sum_{i=0}^{|ch|} \log(n - i) \leq |ch| \log n$, and hence by MDL is the better (i.e., as it is more efficient) way of the two to encode the member nodes of a chain.

2.2.3 ENCODING THE ERRORS

Next, we discuss how we encode the errors made by $\mathbf{M}$ with regard to $\mathbf{A}$, store this information in the *error* matrix $\mathbf{E}$. Many different approaches exist for encoding the errors—amongst which appealing at first glance is to simply identify all node pairs. However, it is important to realize that the more efficient our encoding is, the less spurious "structure" will be discovered.

We hence follow [152] and encode $\mathbf{E}$ in two parts, $\mathbf{E}^+$ and $\mathbf{E}^-$. The former corresponds to the area of $\mathbf{A}$ that M does model, and for which $\mathbf{M}$ includes superfluous edges. Analogue, $\mathbf{E}^-$ consists of the area of $\mathbf{A}$ not modeled by M, for which $\mathbf{M}$ lacks edges. We encode these separately as they are likely to have different error distributions. Note that since we know that near cliques and near bipartite cores are encoded exactly, we ignore these areas in $\mathbf{E}^+$. We encode the edges in $\mathbf{E}^+$ and $\mathbf{E}^-$ similarly to how we encode near-cliques, and have

$$L(\mathbf{E}^+) = \log(|\mathbf{E}^+|) + ||\mathbf{E}^+||l_1 + ||\mathbf{E}^+||'l_0$$

$$L(\mathbf{E}^-) = \underbrace{\log(|\mathbf{E}^-|)}_{\text{\# of edges}} + \underbrace{||\mathbf{E}^-||l_1 + ||\mathbf{E}^-||'l_0}_{\text{edges}} \, .$$

That is, we first encode the number of 1s in $\mathbf{E}^+$ (respectively $\mathbf{E}^-$), after which we transmit the 1s and 0s using optimal prefix codes of length l_1 and l_0. We choose to use prefix codes over a binomial for practical reasons, as prefix codes allow us to easily and efficiently calculate accurate local gain estimates in our algorithm, without sacrificing much encoding efficiency (typically < 1 bit in practice).

Size of the Search Space. Clearly, for a graph of n nodes, the search space $\mathcal{M}$ we have to consider for solving the Minimum Graph Description Problem is enormous, as it consists of *all* possible permutations of the collection $\mathcal{C}$ of *all* possible structures over the vocabulary Ω. Unfortunately, it does not exhibit trivial structure, such as (weak) (anti)monotonicity, that we could exploit for efficient search. Further, Miettinen and Vreeken [153] showed that for a directed graph finding the MDL optimal model of only full-cliques is NP-hard. Hence, we resort to heuristics.

2.3 VOG: VOCABULARY-BASED SUMMARIZATION OF GRAPHS

Now that we have the arsenal of graph encoding based on the vocabulary of structure types, Ω, we move on to the next two key ingredients: finding good candidate structures, i.e., instantiating $\mathcal{C}$, and then mining informative graph summaries, i.e., finding the best model M. An illustration of the algorithm is given in Figure 2.3. The pseudocode of VoG is given in Algorithm 2.1, and

the code is available for research purposes at `https://github.com/GemsLab/VoG_Graph_S`
`ummarization.git`.

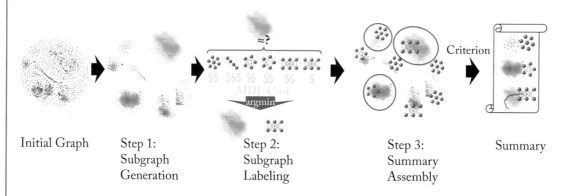

Figure 2.3: Illustration of VoG step-by-step.

2.3.1   SUBGRAPH GENERATION

Any combination of clustering and community detection algorithms can be used to decompose
the graph into subgraphs, which need not be disjoint. These techniques include, but are not limited to *Cross-asssociations* [41], *Subdue* [50], SLASHBURN [139], *Eigenspokes* [176], and *METIS*
[109].

2.3.2   SUBGRAPH LABELING

Given a subgraph from the set of clusters or communities discovered in the previous step, we
search for the structure $x \in \Omega$ that best characterizes it, with no or some errors (e.g., perfect
clique, or clique with some missing edges, encoded as error). A subgraph is labeled with the
structure that, in MDL terms, best approximates the subgraph. To this end, we encode the
subgraph as each of the six candidate vocabulary structures, and choose the structure that has
the lowest encoding cost.

    Let m^* be the graph model with only one subgraph encoded as structure $\in \Omega$ (e.g., clique)
and the additional edges included in the error matrix. For reasons of efficiency, instead of calculating the full cost $L(G, m^*)$ as the encoding cost of each subgraph representation, we estimate
the *local* encoding cost $L(m^*) + L(\mathbf{E}_{m^*}^+) + L(\mathbf{E}_{m^*}^-)$, where $\mathbf{E}_{m^*}^+$ and $\mathbf{E}_{m^*}^-$ encode the incorrectly
modeled, and unmodeled edges, respectively (Section 2.2). The challenge of the step is to ef-
ficiently identify the role of each node in the subgraph (e.g., hub/spoke in a star, member of
set A or B in a near-bipartite core, order of nodes in chain) for the MDL representation. We
elaborate on each structure next.

- **Clique:** This representation is straightforward, as all the nodes have the same structural role. All the nodes are members of the clique or the near-clique. For the full clique, the missing edges are stored in a *local* error matrix, $\mathbf{E}_{fc}$, in order to obtain an estimate of the global encoding cost $L(fc) + L(\mathbf{E}_{fc}^+) + L(\mathbf{E}_{fc}^-)$. For near-cliques we ignore $\mathbf{E}_{nc}$, and, so, the encoding cost is $L(nc)$.

- **Star:** Representing a given subgraph as a near-star is straightforward as well. We find the highest-degree node (in case of a tie, we choose one randomly), and set it to be the hub of the star, and identify the rest of the nodes as the peripheral nodes—which are also referred to as spokes. The additional or missing edges are stored in the local Error matrix, $\mathbf{E}_{st}$. The MDL cost of this encoding is computed as $L(st) + L(\mathbf{E}_{st}^+) + L(\mathbf{E}_{st}^-)$.

- **Bipartite core:** In this case, the problem of identifying the role of each node reduces to finding the maximum bipartite graph, which is known as the max-cut problem, and is NP-hard. The need of a scalable graph summarization algorithm makes us resort to approximation algorithms. In particular, finding an approximation to the maximum bipartite graph can be reduced to semi-supervised classification. We consider two classes which correspond to the two node sets, A and B, of the bipartite graph, and the prior knowledge is that the highest-degree node belongs to A, and its neighbors to B. To propagate these classes/labels, we employ Fast Belief Propagation (FaBP in Chapter 3 and [125]) assuming heterophily (i.e., connected nodes belong to different classes). For near-bipartite cores $L(\mathbf{E}_{nb}^+)$ is omitted.

- **Chain:** Representing the subgraph as a chain reduces to finding the longest path in it, which is also NP-hard. We, therefore, employ the following heuristic. Initially, we pick a node of the subgraph at random, and find its furthest node using BFS (temporary start). Starting from the latter and by using BFS again, we find the subsequent furthest node (temporary end). We then extend the chain by local search. Specifically, we consider the subgraph from which all the nodes that already belong to the chain, except for its endpoints, are removed. Then, starting from the endpoints we employ BFS again. If new nodes are found during this step, they are added in the chain (rendering it a near-chain with few loops). The nodes of the subgraph that are not members of this chain are encoded as error in $\mathbf{E}_{ch}$.

After representing the subgraph as each of the vocabulary structures x, we employ MDL to choose the representation with the minimum (local) encoding cost, and add the structure to the candidate set, $\mathcal{C}$. Finally, we associate the candidate structure with its encoding benefit: the savings in bits for encoding the subgraph by the minimum-cost structure type, instead of leaving its edges unmodeled and including them in the error matrix.

Algorithm 2.1 VoG

Input : graph G
Output : graph summary M and its encoding cost

1: **Step 1: Subgraph Generation.** Generate candidate—possibly overlapping—subgraphs using one or more graph decomposition methods.
2: **Step 2: Subgraph Labeling.** Characterize the type of each subgraph $x \in \Omega$ using MDL, identify the type x as the one that minimizes the local encoding cost. Populate the candidate set $\mathcal{C}$ accordingly.
3: **Step 3: Summary Assembly.** Use the heuristics PLAIN, TOP10, TOP100, GREEDY'NFORGET (Section 2.3.3) to select a non-redundant subset from the candidate structures to instantiate the graph model M. Pick the heuristic with the smallest model description.

2.3.3 SUMMARY ASSEMBLY

Given a set of candidate structures, $\mathcal{C}$, how can we efficiently induce the model M that is the best graph summary? The exact selection algorithm, which considers all the possible ordered combinations of the candidate structures and chooses the one that minimizes the cost, is combinatorial, and cannot be applied to any non-trivial candidate set. Thus, we need heuristics that will give a fast, approximate solution to the description problem. To reduce the search space of all possible permutations, we attach a quality measure to each candidate structure, and consider them in order of decreasing quality. The measure that we use is the encoding benefit of the subgraph, which, as mentioned before, is the number of bits that are saved by encoding the subgraph as structure x instead of noise. Our constituent heuristics are as follows.

- PLAIN: The baseline approach gives all the candidate structures as graph summary, i.e., $M = \mathcal{C}$.
- TOP-K: Selects the top-k candidate structures as sorted according to decreasing quality.
- GREEDY'NFORGET (GNF): Considers each structure in $\mathcal{C}$ sequentially, sorted by descending quality, and iteratively includes each in M: as long as the total encoded cost of the graph does not increase, keeps the structure in M, otherwise it removes it. GREEDY'NFORGET continues this process until all the structures in $\mathcal{C}$ have been considered. This heuristic is more computationally demanding than the plain or top-k heuristics, but still handles large sets of candidate structures efficiently.

VoG employs all the heuristics and by MDL picks the overall best graph summarization, or equivalently, the summarization with the minimum description cost.

2.3.4   TOY EXAMPLE

To illustrate how VoG works, we give an example on a toy graph. We apply VoG on the synthetic Caveman graph of 841 nodes and 7,547 edges which, as shown in Figure 2.4, consists of two cliques separated by two stars. The leftmost and rightmost cliques consist of 42 and 110 nodes, respectively; the big star (2nd structure) has 800 nodes, and the small star (3rd structure) 91 nodes. Here is how VoG works step-by-step.

- **Step 1:** The *raw* output of the decomposition algorithm (SLASHBURN [139]) consists of the subgraphs corresponding to the stars, the full left-hand and right-hand cliques, as well as a few subsets of these nodes.
- **Step 2:** Through MDL, VoG correctly identifies the type of these structures.
- **Step 3:** Finally, via GREEDY'NFORGET, it automatically finds the four true structures without redundancy, and drops the structures that consist of subsets of nodes.

The corresponding model requires 36% fewer bits than the "empty" model, where the graph edges are encoded as noise.

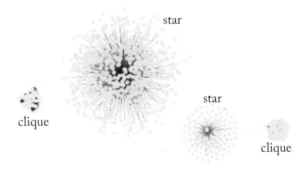

Figure 2.4: Toy graph: VoG saves 36% in space, by successfully discovering the two cliques and two stars that we chained together.

2.3.5   TIME COMPLEXITY

For a graph $G(\mathcal{V}, \mathcal{E})$ of $n = |\mathcal{V}|$ nodes and $m = |\mathcal{E}|$ edges, the time complexity of VoG depends on the runtime complexity of the algorithms that compose it, namely the decomposition algorithm, the subgraph labeling, the encoding scheme $L(G, M)$ of the model, and the structure selection (summary assembly).

For the *decomposition* of the graph, we use SLASHBURN which is near-linear on the number of edges of real graphs [139]. The *subgraph labeling* algorithms in Section 2.3.2 are carefully designed to be linear on the number of edges of the input subgraph.

When there is no overlap between the structures in M, the complexity of calculating the *encoding scheme* $L(G, M)$ is $O(m)$. When there is some overlap, the complexity is bigger: assume that s, t are two structures $\in M$ with overlap, and t has higher quality than s, i.e., t comes before s in the ordered list of structures. Finding how much "new" structure (or area in **A**) s explains relative to t costs $O(|M|^2)$. Thus, in the case of overlapping subgraphs, the complexity of computing the encoding scheme is $O(|M|^2 + m)$. As typically $|M| \ll m$, in practice we have $O(m)$.

As far as the *selection method* is concerned, the Top-κ heuristic that we propose has complexity $O(k)$. The Greedy'nForget heuristic has runtime $O(|\mathcal{C}| \times o \times m)$, where $|\mathcal{C}|$ is the number of structures identified by VoG, and o the time complexity of $L(G, M)$.

2.4   EMPIRICAL RESULTS

In this section, we aim to answer the following questions.

Q1. Are the real graphs structured, or random and noisy? If the graphs are structured, can their structures be discovered under noise?

Q2. What structures do the graph summaries contain, and how can they be used for understanding?

Q3. Is VoG scalable and able to efficiently summarize large graphs?

The graphs we use in the experiments along with their descriptions are summarized in Table 2.2. Liancourt-Rocks is a co-editor graph on a controversial Wikipedia article about the island Liancourt Rocks, where the nodes are users, and the edges mean that they edited the same sentence. Chocolate is a co-editor graph on the "Chocolate" article. The descriptions of the other datasets are given in Table 2.2.

Table 2.2: VoG: Summary of graphs used

Name	Nodes	Edges	Description
Flickr [73]	404,733	2,110,078	Friendship social network
WWW-Barabasi [207]	325,729	1,090,108	WWW in nd.edu
Epinions [207]	75,888	405,740	Trust graph
Enron [63]	80,163	288,364	Enron email
AS-Oregon [19]	13,579	37,448	Router connections
Wikipedia-Liancourt-Rocks	1,005	2,123	Co-edit graph
Wikipedia-Chocolate	2,899	5,467	Co-edit graph

Graph Decomposition. In our experiments, we modify SlashBurn [139], a node reordering algorithm, to generate candidate subgraphs. The reasons we use SlashBurn are (a) it is scalable, and (b) it is designed to handle graphs *without* caveman structure. We note that VoG would only benefit from using the outputs of additional decomposition algorithms.

SLASHBURN is an algorithm that reorders the nodes so that the resulting adjacency matrix has clusters or patches of non-zero elements. The idea is that removing the top high-degree nodes in real-world graphs results in the generation of many small-sized disconnected components (subgraphs), and one giant connected component whose size is significantly smaller compared to the original graph. Specifically, it performs two steps iteratively: (a) it removes top high degree nodes from the original graph; and (b) it reorders the nodes so that the high-degree nodes go to the front, the disconnected components to the back, and the giant connected component (GCC) to the middle. During the next iterations, these steps are performed on the giant connected component. A good node-reordering method will reveal patterns, as well as large empty areas, as shown in Figure 2.5 on the Wikipedia Chocolate network.

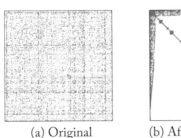 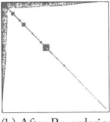

(a) Original   (b) After Re-ordering

Figure 2.5: Adjacency matrix before and after node-ordering on the Wikipedia Chocolate graph. Large empty (and dense) areas appear, aiding the graph decomposition step of VoG and the discovery of candidate structures.

In this work, SLASHBURN is modified to decompose the input graph. In more details, we focus on the first step of the algorithm, which removes the high degree node by "burning" its edges. This step is depicted for a toy graph in Figure 2.6b, where the green dotted line shows which edges were "burnt." Then, the hub with its egonet, which consists of the hub's one hop away neighbors and the connections between them, form the first candidate structures. Moreover, the connected components with a size greater or equal to two and smaller than the size of the GCC, consist of additional candidate structures (see Figure 2.6c). In the next iteration, the same procedure is applied to the giant connected component, yielding this way a set of candidate structures. We use MDL to determine the best-fitting type per discovered candidate structure.

2.4.1   QUANTITATIVE ANALYSIS

In this section we apply VoG to the real datasets of Table 2.2, and evaluate the achieved description cost, and edge coverage, which are indicators of the discovered structures. The evaluation is done in terms of savings with respect to the base encoding (ORIGINAL) of the adjacency matrix of a graph with an empty model M. Moreover, by using synthetic datasets, we evaluate the ability of VoG to discover the ground truth graph structures under the presence of noise. Finally, we

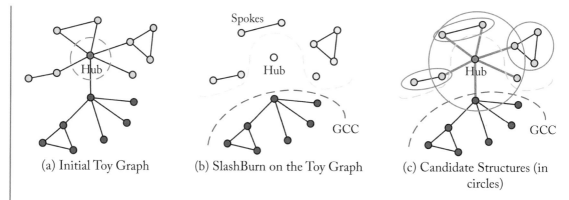

(a) Initial Toy Graph   (b) SlashBurn on the Toy Graph   (c) Candidate Structures (in circles)

Figure 2.6: Illustration of the graph decomposition and the generation of the candidate structures.

discuss the selection of the vocabulary for summarization, and compare the different selections quantitatively in terms of the encoding cost of the summaries that they generate.

Description Cost

Although we refer to the description cost of the summarization techniques, we note that compression itself is *not* our goal, but our *means* for identifying the structures important for graph understanding or attention routing.[3] This is also why it does not make sense to compare VoG with standard matrix compression techniques: whereas VoG has the goal of describing a graph with simple and easy-to-understand structures, specialized algorithms may exploit any statistical correlations to save bits.

We compare three summarization approaches: (a) ORIGINAL: The whole adjacency matrix is encoded as if it contains no structure; that is, $M = \emptyset$, and $\mathbf{A}$ is encoded through $L(\mathbf{E}^-)$; (b) SB+nc: All the subgraphs extracted by our method (first step of Algorithm 2.1 using our proposed variant of SLASHBURN) are encoded as near-cliques; and (c) VoG: Our proposed summarization algorithm with the three selection heuristics (PLAIN, TOP10 and TOP100, GREEDY'NFORGET[4]).

We ignore very small structures; the candidate set $\mathcal{C}$ includes subgraphs with at least 10 nodes, except for the Wikipedia graphs where the size threshold is set to 3 nodes. Among the summaries obtained by the different heuristics, we choose the one that yields the smallest description length.

---

[3]High compression ratios are exactly a sign that many redundancies (i.e., patterns) that can be explained in simple terms (i.e., structures) were discovered.

[4]By carefully designing the GREEDY'NFORGET heuristic to exploit memoization, we are able to efficiently compute the best structures within the candidate set. Although restricting our search space to a small number of candidate structures—ranked in decreasing order of quality—can yield faster results, we report results on the whole search space.

Table 2.3 presents the summarization cost of each technique with respect to the cost of the ORIGINAL approach, as well as the fraction of the edges that remains unexplained. Specifically, the first column, ORIGINAL, presents the cost, in bits, of encoding the adjacency matrix with an empty model M. The second column, SB+nc, presents the relative number of bits needed to describe the structures discovered by SLASHBURN as near-cliques. Then, for different VoG heuristics we show the relative number of bits needed to describe the adjacency matrix. In the last four columns, we give the fraction of edges that are *not* explained by the structures in the model, M, that each heuristic finds. The lowest description cost is in bold.

Table 2.3: [Lower is better.] Quantitative comparison of baseline methods and VoG with different summarization heuristics. The first column, ORIGINAL, presents the cost, in bits, of encoding the adjacency matrix with an empty model M. The other columns give the relative number of bits needed to describe the adjacency matrix.

Graph	Original (bits)	SB + nc (% bits)	VoG							
			Compression				Unexplained Edges			
			Plain	Top 10	Top 100	GnF	Plain	Top 10	Top 100	GnF
Flickr	35,210,972	92%	**81%**	99%	97%	95%	4%	72%	39%	36%
WWW–Barabasi	18,546,330	94%	**81%**	98%	96%	85%	3%	62%	51%	38%
Epinions	5,775,964	128%	82%	98%	95%	**81%**	6%	65%	46%	14%
Enron	4,292,729	121%	**75%**	98%	93%	**75%**	2%	77%	46%	6%
AS–Oregon	475,912	126%	72%	87%	79%	**71%**	4%	59%	25%	12%
Chocolate	60,310	127%	96%	96%	93%	**88%**	4%	70%	35%	27%
Lian-court-Rocks	19,833	138%	98%	94%	96%	**87%**	5%	51%	12%	31%

The lower the ratios (i.e., the lower the obtained description length), the more structure is identified. For example, VoG-PLAIN describes Flickr with only 81% of the bits of the ORIGINAL approach, and explains all but 4% of the edges, which means that 4% of the edges are not encoded by the structures in M.

Observation 2.3   Real graphs do have structure; VoG, with or without structure selection, achieves better compression than the ORIGINAL approach that assumes no structure, as well as the SB+nc approach that encodes all the subgraphs as near-cliques.

We observe that the SB+nc approach often requires more bits than the ORIGINAL approach, which assumes no structure. This is due to the fact that the discovered structures often overlap and, thus, some nodes are encoded multiple times. Moreover, a near-clique is not the optimal model for all the extracted structures; if it were, VoG-PLAIN (which is more expressive and allows more models, such as full clique, bipartite core, star, and chain) would have resulted in higher encoding cost than the SB+nc method.

GREEDY'NFORGET finds models M with fewer structures than PLAIN and TOP100—which is important for graph understanding and guiding attention to few structures–, and often obtains (much) more succinct graph descriptions. This is due to its ability to identify structures that are informative *with regard to what it already knows*. In other words, structures that highly overlap with ones already selected into M will be much less rewarded than structures that explain unexplored parts of the graph.

Discovering Structures under Noise

In this section we evaluate whether VoG is able to detect the underlying graph structures under noise. We start with the Caveman graph described in Section 2.3.4 (original graph), and generate noisy instances by reverting

$$\epsilon = \{0.001, 0.005, 0.01, 0.05, 0.10, 0.15, 0.2, 0.25, 0.3, 0.35\}$$

of the original graph's edges ($\mathcal{E}_{orig}$). For example, at noise level $\epsilon = 0.001$, we randomly pick $0.001|\mathcal{E}_{orig}|$ pairs of nodes. If an edge existed between two selected nodes in the original graph, we remove it in the noisy instance; otherwise, we add it. To evaluate our method's ability to detect the underlying structures, we treat the building blocks of the original graph as ground truth structures, and compute the precision and recall of VoG-GREEDY'NFORGET at different levels of noise. We define the precision as:

$$\text{precision} = \frac{\text{\# of relevant retrieved structures}}{\text{\# of retrieved structures}},$$

where we consider a structure relevant if it (i) overlaps with a ground truth structure, and (ii) has the same, or similar (full and near-clique, or full and near-bipartite core) connectivity pattern to the overlapping ground truth structure.

We define recall as:

$$\text{recall} = \frac{\text{\# of retrieved ground truth structures}}{\text{\# of relevant ground truth structures}},$$

where we consider a ground truth structure retrieved if VoG returned at least one overlapping structure with the same or similar connectivity pattern.

In addition to the precision and recall, we also define the "weighted precision," which penalizes retrieved VoG structures that partially overlap with the ground truth structures.

$$\text{weighted-precision} = \frac{\sum \text{node-overlap of relevant retrieved structures}}{\text{\# of retrieved structures}},$$

where the numerator is the sum of the node overlap of relevant retrieved structures and the corresponding ground truth structures.

In our experiment, we generate ten noisy instances of the original graph at each noise level ϵ. In Figure 2.7, we give the precision, recall, and weighted precision averaged over the 10 graph instances at each level of noise. The error bars in the plot correspond to the standard deviation of the quantities. In addition to the accuracy metrics, we provide the average number of retrieved structures per noise level in Figure 2.8.

We observe that at all noise levels, VoG has high precision and recall (above 0.85 and 0.75, respectively). The weighted precision decreases as the noise increases, but it remains high at low levels of noise (< 0.05).

Observation 2.4   VoG routes attention to the ground truth structures even under the presence of noise.

The high precision and recall of VoG at levels of noise greater than 5% are due to the big number of structures that are retrieved (≈ 20). As evidenced by the weighted precision, the retrieved structures are relevant to the ground truth structures (they are counted as "hits" by the definition of precision and recall), but they do not recover the ground truth structures perfectly[5] leading to lower levels of weighted precision. On the other hand, for noise below 5%, we observe a small drop in the precision and recall. Although the retrieved and ground truth structures are almost equal in number, and have high node overlap, they do not always have the same connectivity patterns (e.g., a star matches to a near-bipartite core), which leads to a slight decrease in the precision and recall of VoG.

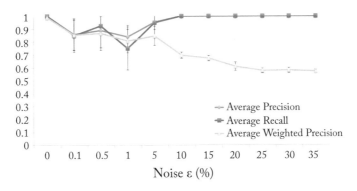

Figure 2.7: VoG routes attention to the ground truth structures even under the presence of noise. We show the precision, recall, and weighted-precision of VoG at different levels of noise.

2.4.2   QUALITATIVE ANALYSIS

In this section, we showcase how to use VoG and interpret the graph summaries that it outputs.

---

[5]The overlap of the retrieved structures and the ground truth structures is often < 1.

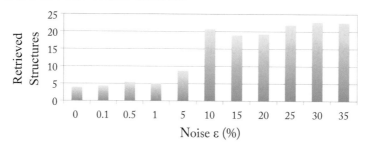

Figure 2.8: Number of structures retrieved under the presence of noise. As the noise increases and more loosely connected components are created, VoG retrieves more structures.

Graph Summaries

How well does VoG summarize real graphs? Which are the most frequent structures? Table 2.4 shows the summarization results of VoG for different structure selection techniques.

Table 2.4: Summarization of graphs by VoG. The most frequent structures are the stars and near-bipartite cores. We provide the frequency of each structure type: "st" for star, "nb" for near-bipartite cores, "fc" for full cliques, "fb" for full bipartite-cores, "ch" for chains, and "nc" for near-cliques.

Graph	Plain						Top 10		Top 100				Greedy'nForget			
	st	nb	fc	fb	ch	nc	st	nb	st	nb	fb	ch	st	nb	fb	ch
Flickr	24,385	3,750	281	9	–	3	10	–	99	1	–	–	415	–	–	1
WWW-Barabasi	10,027	1,684	487	120	26	–	9	1	83	14	3	–	4,177	161	328	85
Epinions	5,204	528	13	–	–	–	9	1	99	1	–	–	2,644	–	8	–
Enron	3,171	178	3	11	–	–	9	1	99	1	–	–	1,810	–	2	2
AS-Oregon	489	85	–	4	–	–	10	–	93	6	1	–	399	–	–	–
Chocolate	170	58	–	–	17	–	9	1	87	10	–	3	101	–	–	–
Lian-court-Rocks	73	21	–	1	22	–	8	2	66	17	1	16	39	–	–	–

Observation 2.5   The summaries of all the selection heuristics consist mainly of stars, followed by near-bipartite cores. In some graphs, like Flickr and WWW-Barabasi, there are a significant number of full cliques.

From Table 2.4 we also observe that Greedy'nForget drops uninteresting structures, and reduces the graph summary. Effectively, it filters out the structures that explain edges already explained by structures in model M.

How often do we find perfect cliques, bipartite cores, etc., in real graphs? To each structure, we attach a quality score that quantifies how close the structure that VoG discovered (e.g., a near-bipartite core) is to the "perfect" structure consisting of the same nodes (e.g., perfect bipartite score on the same nodes, without any errors). We define the quality score of structure s as:

$$quality(s) = \frac{\text{encoding cost of error-free structure} \in \Omega}{\text{encoding cost of discovered structure}}.$$

The quality score takes values between 0 and 1. A quality score that tends to 0 corresponds to a structure that deviates significantly from the "perfect" structure, while 1 means that the discovered structure is perfect (e.g., error-free star). Table 2.5 gives the average quality of the structures discovered by VoG in real datasets.

Table 2.5: Quality of the structures discovered by VoG. For each structure type, we provide the average (and standard deviation) of the quality of the discovered structures.

Graph	st	cnb	fc	fb	ch	nc
WWW-Barabasi	0.78 (0.25)	0.77 (0.22)	0.55 (0.17)	0.51 (0.42)	1 (0)	-
Epinions	0.66 (0.27)	0.82 (0.15)	0.50 (0.08)	-	-	-
Enron	0.62 (0.65)	0.85 (0.19)	0.53 (0.02)	1 (0)	-	-
AS-Oregon	0.65 (0.30)	0.84 (0.18)	-	1 (0)	-	-
Chocolate	0.75 (0.20)	0.89 (0.19)	-	-	1 (0)	-
Liancourt-Rocks	0.75 (0.26)	0.94 (0.14)	-	1 (0)	1 (0)	-

By leveraging the MDL principle, VoG can discover not only exact structures, but also approximate structures that have some erroneous edges. In the real datasets that we studied, the chains that VoG discovers do not have any missing or additional edges; this is probably due to the small size of the chains (4 nodes long, on average, for Chocolate and Liancourt-Rocks, and 20 nodes long for WWW-Barabasi). The near-bipartite cores, and stars are also of high quality (at least 0.77 and 0.66, respectively). Finally, the discovered cliques are almost half-full, as evidenced by the 0.50–0.55 quality score.

In order to gain a better understanding of the structures that VoG finds, in Figure 2.9 we give the size distributions of the most frequent structures in the Flickr social network.

Observation 2.6   The size distribution of the stars and near-bipartite cores follows a power law.

The distribution of the size of the full cliques in Flickr follows a power law as well, while the distributions of the full cliques and bipartite cores in WWW-Barabasi do not show any clear pattern. In Figures 2.9 and 2.10, we denote with blue crosses the size distribution of the structures discovered by VoG-PLAIN, and with red circles the size distribution for the structures found by VoG with the TOP100 heuristic.

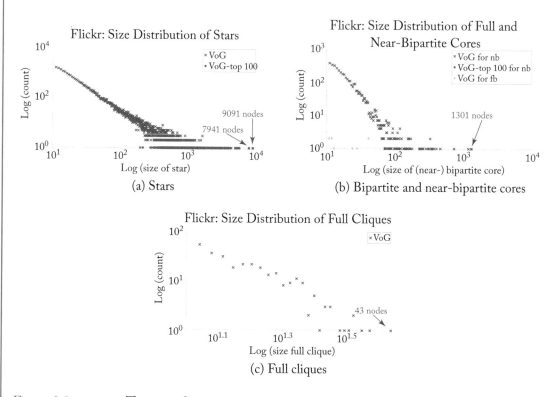

Figure 2.9: Flickr: The size of stars, near-bipartite cores and full cliques follows the power law distribution. Distribution of the size of the structures by VoG (blue crosses) and VoG-TOP100 (red circles) that are the most informative from an information-theoretic point of view.

Graph Understanding

Are the "important" structures found by VoG semantically meaningful? For sense-making, we analyze the discovered subgraphs in three non-anonymized real datasets: Wikipedia–Liancourt-Rocks, Wikipedia–Chocolate, and Enron.

Wikipedia—Liancourt-Rocks   Figures 2.1 and 2.11a–b illustrate the original and VoG-based visualization of the Liancourt-Rocks graph. The VoG-TOP10 summary consists of eight stars and two near-bipartite cores (see also Table 2.4). To visualize the graph we leveraged the structures to which VoG draws attention by using Cytoscape[6]: For Figure 2.11a, we applied the

[6]http://www.cytoscape.org/

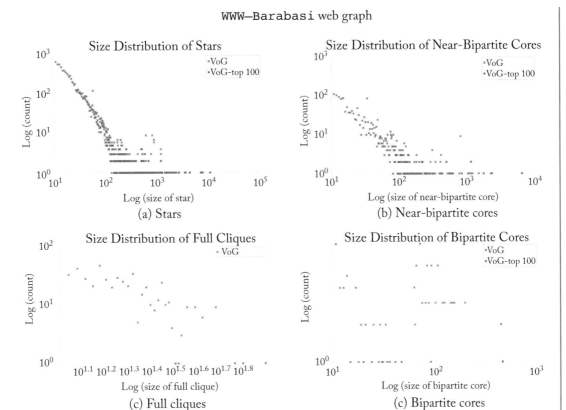

Figure 2.10: The distribution of the size of the structures (starts, near-bipartite cores) that are the most "interesting" from the MDL point of view, follow the power law distribution both in the WWW-Barabasi web graph. The distribution of structures discovered by VoG and VoG-TOP100 are denoted by blue crosses and red circles, respectively.

spring-embedded layout to the Wikipedia graph, and then highlighted the centers of the eight stars that VoG discovered by providing the list of their corresponding IDs. For Figure 2.11b, we input the list of nodes belonging to one side of the most "important" bipartite core that VoG discovered, selected, and dragged the corresponding nodes to the left top corner, and applied the circular layout to them. The second most important bipartite core is visualized in the same way in Figure 2.1d.

The eight star configurations correspond mainly to administrators, such as "Future_Perfect_at_sunrise," who do many minor edits in various parts of the article and also revert vandalisms. The most interesting structures VoG identifies are the near-bipartite cores, which reflect: (i) the conflict between the two parties about the territorial rights to the

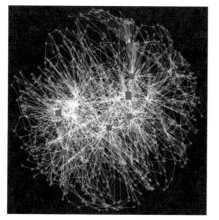

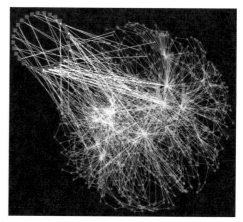

(a) VoG: The eight most "important" stars
(their centers denoted with red rectangles).

(b) VoG: The most "important" bipartite graph
(node set A denoted by the circle of red points).

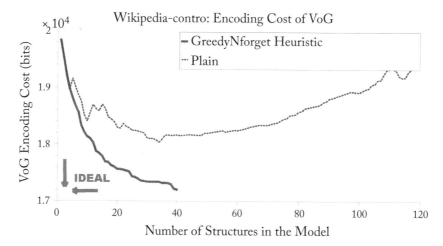

(c) Effectiveness of GREEDY'NFORGET (in red). Encoding cost of VoG vs. number of structures in the model, M.

Figure 2.11: The VoG summary of the Liancourt-Rocks graph, and effectiveness of the GREEDY'NFORGET heuristic. In (c), GREEDY'NFORGET leads to better encoding costs and smaller summaries (here only 40 are chosen) than PLAIN ($\sim$ 120 structures).

island (Japan vs. South Korea), and (ii) an "edit war" between vandals and administrators or loyal Wikipedia users.

In Figure 2.11c, the encoding cost of VoG is given as a function of the selected structures. The dotted blue line corresponds to the cost of the PLAIN encoding, where the structures are added sequentially in the model M, in decreasing order of quality (local encoding benefit). The solid red line maps to the cost of the GREEDY'NFORGET heuristic. Given that the goal is to summarize the graph in the most succinct way, and at the same time achieve low encoding cost, GREEDY'NFORGET is effective. Finally, in Figure 2.12 we consider the models M with increasing number of structures (in decreasing quality order) and show the number of edges that each one explains. Specifically, Figure 2.12a refers to the models that consist of ordered subsets of all the structures (~120) that VoG-PLAIN discovered and Figure 2.12b refers to models incorporating ordered subsets of the ~35 structures kept by VoG-GREEDY'NFORGET. We note that the slope in Figure 2.12a is steep for the first ~ 35 structures, and then increases with small rate, which means that the new structures that are added in the model M explain few new edges (diminishing returns). On the other hand, the edges explained by the structures discovered by VoG-GREEDY'NFORGET increase with higher rate, which signifies that VoG-GREEDY'NFORGET drops uninteresting structures that explain edges that are already explained by structures in model M.

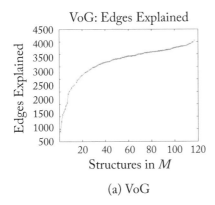

(a) VoG

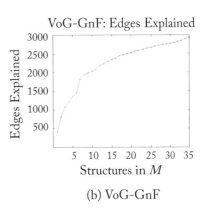

(b) VoG-GnF

Figure 2.12: Wikipedia-Liancourt-Rocks: VoG-GnF successfully drops uninteresting structures that explain edges already explained by structures in model M. Diminishing returns in the edges explained by the new structures that VoG and VoG-GnF select.

Wikipedia—Chocolate   The visualization of Wikipedia Chocolate is similar to the visualization of the Liancourt-Rocks and is given in Figure 2.13 for completeness. As shown in Table 2.4, the TOP10 summary of Chocolate contains nine stars and one near-bipartite core. The center of the highest-ranked star corresponds to "Chobot," a Wikipedia bot that fixes interlanguage links, and thus touches several, possibly unrelated parts of a page. The hubs (centers) of other stars correspond to administrators, who do many minor edits, as well as other heavy contributors. The near-bipartite core captures the interactions between possible vandals and administrators (or Wikipedia contributors) who were reverting each other's edits resulting in temporary (semi-) protection of the web page. Figure 2.13c illustrates the effectiveness of

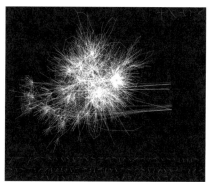

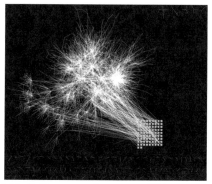

(a) VoG: The nine most "important" stars (the hubs of the stars denoted by the cyan points).

(b) VoG: The most "important" bipartite graph (node set A denoted by the circle of rectangle of cyan points).

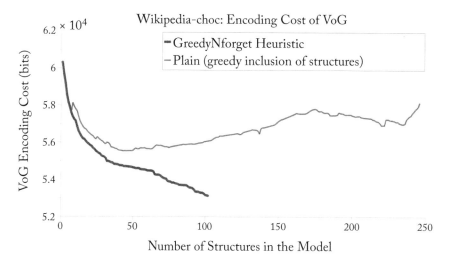

(c) Effectiveness of GREEDY'NFORGET (in red). Encoding cost of VoG vs. the number of structures in the model, M.

Figure 2.13: **VoG:** summarization of the structures of the Wikipedia `Chocolate` graph. (a)–(b): The top-10 structures of the summary. (c): The GREEDY'NFORGET heuristic (red line) reduces the encoding cost by keeping approximately the 100 most important of the 250 identified structures.

the GREEDY'NFORGET heuristic when applied to the Chocolate article. The GREEDY'NFORGET heuristic (red line) reduces the encoding cost by keeping approximately the 100 most important of the 250 identified structures (x-axis). The blue line corresponds to the encoding cost (y-axis) of greedily adding the identified structures in decreasing order of encoding benefit.

2.4.3 SCALABILITY

In Figure 2.14, we present the runtime of VoG with respect to the number of edges in the input graph. For this purpose, we induce subgraphs of the Notre Dame dataset (WWW-Barabasi) for which we give the dimensions in Table 2.6. We ran the experiments on an Intel(R) Xeon(R) CPU 5160 at 3.00 GHz, with 16 GB memory. The structure identification is implemented in Matlab, while the selection process in Python.

Observation 2.7   All the steps of VoG are designed to be scalable. Figure 2.14 shows the complexity is $O(m)$, i.e., VoG is near-linear on the number of edges of the input graph.

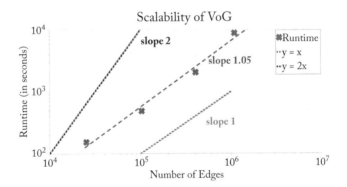

Figure 2.14: VoG is near-linear on the number of edges. Runtime, in seconds, of VoG (PLAIN) vs. number of edges in graph. For reference we show the linear and quadratic slopes.

Table 2.6: Scalability: Induced subgraphs of WWW-Barabasi

Name	Nodes	Edges
WWW-Barabasi-50k	49,780	50,624
WWW-Barabasi-100k	99,854	205,432
WWW-Barabasi-200k	200,155	810,950
WWW-Barabasi-300k	325,729	1,090,108

2.5   DISCUSSION

The experiments show that VoG successfully solves an important open problem in graph under-standing: how to find a succinct summary for a large graph. In this section we discuss some of our design decisions, as well as the advantages and limitations of VoG.

Why does VoG use the chosen vocabulary structures consisting of stars, (near) cliques, (near) bipartite cores and chains, and not other structures?   The reason for choosing these and not other structures is that these structures appear very often, in tens of real graphs, (e.g., in patent citation networks, phonecall networks, in the Netflix recommendation system, etc.), while they also have semantic meaning, such as factions or popular entities. Moreover, these graph structures are well-known and conceptually simple, making the summaries that VoG discovers easily interpretable.

It is possible that a graph does not contain the vocabulary terms we have predefined, but has more complex structures. However, this does not mean that the summary that VoG would generate will be empty. A core property of MDL is that it identifies the model in a model class that best describes your graph regardless of whether the *true* model is in that class. Thus, VoG will return the model that gives the most succinct description of the input graph in the vocabulary terms at hand: our model class. In this case, VoG will give a more crude description of the graph structure than would be ideal—it will have to spend more bits than ideal, which means we are guarded against overfitting. For the theory behind MDL for model selection, see [88].

Can VoG handle the case where a new structure (e.g., "loops") proves to be frequent in real graphs?   If a new structure is frequent in real graphs, or other structures are important for specific applications, VoG can easily be extended to handle new vocabulary terms. The key in-sight for the vocabulary term encodings in Section 2.2 is to encode the necessary information as succinctly as possible. In fact, as by MDL we can straightforwardly compare two or more model classes, MDL will immediately tell us whether a vocabulary set $\mathcal{V}_1$ is better than a vocabulary set $\mathcal{V}_2$: the one that gives the best compression cost for the graph wins.

On the other hand, if we already know that certain nodes form a clique, star, etc., it is trivial to adapt VoG to use these structures as the base model M of the graph (as opposed to the empty model). When describing the graph, VoG will then only report those structures that best describe the remainder of the graph.

Why does VoG fix the vocabulary *a priori* instead of automatically determining the most ap-propriate vocabulary for a given graph?   An alternative approach to fixing the vocabulary would be to automatically determine the "right" vocabulary for a given graph by doing frequent graph mining. We did not pursue this approach because of scalability and interpretability. For a vocabulary term to be useful, it needs to be easily understood by the user. This also relates to why we define our own encoding and optimization algorithm, instead of using off-the-shelf general-purpose compressors based on Lempel-Ziv (such as gzip) or statistical compressors (such as

PPMZ); these provide state-of-the-art compression by exploiting complex statistical properties of the data, making their models very complex to understand. Local structure-based summaries, on the other hand, are much more easily understood. Frequent patterns have been proven to be interpretable and powerful building blocks for data summarization [121, 205, 217]. While a powerful technique, spotting frequent subgraphs has the notoriously expensive subgraph isomorphism problem in the inner loop. This aside, published algorithms on discovering frequent subgraphs (e.g., [225]), are not applicable here, since they expect the nodes to have labels (e.g., carbon atom, oxygen atom, etc.) whereas we focus on large unlabeled graphs.

Can VoG be extended such that it can take specific edge distributions into account and only report structures that stand out from such a distribution?   In this work we aim to assume as little as necessary for the edge distribution, such that VoG is both parameter-free and non-parametric at its core. However, it is possible to use a specific edge distribution. As long as we can calculate the probability of an adjacency matrix, $P(\mathbf{E})$, we can trivially define $L(\mathbf{E}) = -\log P(\mathbf{E})$. Thus, for instance, if we were to consider a distribution with a higher clustering coefficient (that is, dense areas are more likely), the cost for having a dense area in $\mathbf{E}$ will be relatively low, and hence VoG will only report structures that stand out from this (assumed) background distribution. Recent work by Araujo et al. [17] explores discovering communities that exhibit a different hyperbolic—power-law degree distributed—connectivity than the background distribution. It will be interesting to extend VoG with hyperbolic distributions for both subgraphs, as well as for encoding the error matrix.

Why use SlashBurn for the graph decomposition?   To generate candidate subgraphs, we use SLASHBURN because it is scalable, and designed to handle graphs *without* caveman structure. However, *any* graph decomposition method could be used instead, or even better a combination of decomposition methods could be applied. We conjecture that the more graph decomposition methods provide the candidate structures for VoG, the better the resulting summary will be. Essentially, there is no correct graph partitioning technique, since each one of them works by optimizing a different goal. MDL, which is an indispensable component of VoG, will be able to discover the best structures among the set of candidate structures.

Can VoG give high-level insight in how the structures in a summary are connected?   Although VoG does not explicitly encode the linkage between the discovered structures, it can give high-level insight into how the structures in a summary are connected. From the design point of view, we allow nodes to participate in multiple structures, and such nodes implicitly "link" two structures. For example, a node can be part of a clique, as well as the starting-point of a chain, therefore "linking" the clique and the chain. The linkage structure of the summary can hence be trivially extracted by inspecting whether the node sets of structures in the summary overlap. It may depend on the task whether one prefers to show the high level linkage of the structure, or give the details per structure in the summary.

2.6 RELATED WORK

Work related to VoG comprises Minimum Description Length-based approaches, as well as graph compression and summarization, partitioning, and visualization.

Minimum Description Length (MDL) Faloutsos and Megalooikonomou [63] argue that since many data mining problems are related to summarization and pattern discovery, they are intrinsically related to Kolmogorov complexity. Kolmogorov complexity [138] identifies the shortest lossless algorithmic description of a dataset, and provides sound theoretical foundations for both identifying the optimal model for a dataset, and defining what structure it is. While not computable, it can be practically implemented by the Minimum Description Length principle [88, 183] (lossless compression).

Graph Compression and Summarization A recent survey [141] reviews various graph summarization methods. In summary, [31] studied the compression of web graphs using the lexicographic localities; [46] extended it to the social networks; [16] uses BFS for compression; [149] uses multi-position linearizations for neighborhood queries; [67] encodes edges per triangle using a lossy encoding; and [144] reduce the redundancy around high-degree nodes to speed up pattern matching queries. SLASHBURN [139] performs node reordering and graph compression by exploiting the power-law behavior of real-world graphs. Tian et al. [207] present an attribute-based graph summarization technique with non-overlapping and covering node groups; an automation of this method is given by Zhang et al. [233]. Toivonen et al. [208] use a node structural equivalence-based approach for compressing weighted graphs.

Assuming away the "no good cut" issue, there are countless graph partitioning algorithms: Koopman and Siebes [120, 121] summarize multi-relational data, or heavily attributed graphs. Their method assumes the adjacency matrix is already known, as it aims at describing the node attribute-values using tree-shaped patterns. SUBDUE [50] is a popular frequent subgraph-based summarization scheme. It iteratively replaces the most frequent subgraph in a labeled graph by a meta-node, which allows it to discover small lossy descriptions of labeled graphs. In contrast, we consider unlabeled graphs. Moreover, as our encoding is lossless, by MDL, we can fairly compare radically different models and model classes. Navlakha et al. [158] follow a similar approach to Cook and Holder [50], by iteratively grouping nodes that see high interconnectivity. Their method is hence confined to summarizing a graph in terms of non-overlapping cliques and bipartite cores. In comparison, the work of Miettinen and Vreeken [153] is closer to ours, even though they discuss MDL for Boolean matrix factorization. For directed graphs, such factorizations are in fact summaries in terms of possibly overlapping full cliques. Complementary to these works is the detection of network motifs, i.e., recurring, significant connectivity patterns, which compose complex networks [154, 155, 211, 226]. This line of work focuses on "summarizing" graphs with subgraph frequency statistics as opposed to a smaller network representation.

An alternative way of "compressing" a graph is by sampling nodes or edges from it [96, 132] with the goal of maintaining some properties of the initial graph, such as the degree distribution, the size distribution of connected components, the diameter, or latent properties including the community structure [147] (i.e., the graph sample contains nodes from all the existing communities). Although graph sampling may allow for better visualization [179], unlike VoG, it cannot detect graph structures and may need additional processing in order to make sense of the sample.

With VoG we go beyond a single-term vocabulary, and, importantly, we can detect and reward explicit structure within subgraphs.

Graph Partitioning Chakrabarti et al. [40] proposed the cross-association method, which provides a hard clustering of the nodes into groups, effectively looking for near-cliques. Papadimitriou et al. [168] extended this to hierarchies, again of hard clusters. Rosvall and Bergstrom [186] propose information-theoretic approaches for community detection for hard-clustering the nodes of the graph. With VoG we allow nodes to participate in multiple structures, and can summarize graphs in terms of *how* subgraphs are connected, beyond identifying that they are densely connected.

The blockmodels representation [65] is an alternative approach to partition the nodes (actors) of a graph into groups, and capture the network relations between them. The idea of blockmodels is relevant to our approach, which summarizes a graph in terms of simple structures, and reveals the connections between them. Particularly, the mixed-membership assumption that we make in this section is related to the stochastic blockmodels [4, 107]. These probabilistic models combine global parameters that instantiate dense patches of connectivity (blockmodel) with local parameters that capture nodes belonging to multiple blockmodels. Unlike our method, blockmodels need the number of partitions as input, spot mainly dense patches (such as cliques and bipartite cores, without explicitly characterizing them) in the adjacency matrix, and make a set of statistical assumptions about the interaction patterns between the nodes (generative process). A variety of graph clustering methods, including blockmodels, could be used in the first step of VoG (Algorithm 2.1) in order to generate candidate subgraphs, which would then be ranked, and maybe included in the graph summary. A study of various graph clustering methods in terms of their summarization power is presented in [142].

Graph Visualization VoG enables visualization of large-scale networks by focusing on their important parts. Dunne and Shneiderman [58] introduce the idea of motif simplification to enhance network visualization. Some of these motifs are part of our vocabulary, but VoG also allows for near-structures, which are common in real-world graphs.

In general, most graph visualization techniques focus on anomalous nodes or how to visualize the patterns of the whole graph. Other graph visualization tools include: Apolo [42], in which a user picks a few seeds nodes and Apolo interactively expands their vicinities enabling sense-making this way; OPAvion [5], an anomaly detection system for large graphs that mines graph features on Hadoop, spots anomalies by employing OddBall [7], and lastly interactively visualizes the anomalous nodes via Apolo; scaled density plots to visualize scatter plots [195];

random and density sampling [26] for datasets with thousands of points; and rescaled visualization of spy, distribution and correlation plots of massive graphs [104]. PERSEUS takes a different approach to summarization; it is a large-scale system that enables the comprehensive analysis of large graphs by supporting the coupled summarization of graph properties (computed in a distributed way on Hadoop or Spark) and structures, guiding attention to outliers, and allowing the user to interactively explore normal and anomalous node behaviors in distribution plots and ego-network representations. An extension of this work, EAGLE provides an automated way for summarizing any graph with a set of representative invariant distributions by learning and leveraging the domain knowledge.

Visualization-based graph summarization is also relevant to visual graph analytics, since summarization can enable and support interactive visualization. However, the typical goal of graph summarization is quite different from that of visual graph analytics, which focus mostly on the display layouts, new visualization and interaction techniques etc. Widely used visualization tools, such as Gephi, Cytoscape, and the Javascript D3 library support interactive, real-time exploration of networks and operations such as spatializing, filtering and clustering. These tools can benefit from graph summarization methods that result in smaller network representations or patterns thereof, which can be displayed more easily.

CHAPTER 3

Inference in a Graph

In Chapter 2 we saw how we can summarize a large graph and gain insights into its important and semantically meaningful structures. In this chapter we examine how we can use the network effects to learn about the nodes in the remaining network structures. In contrast to the previous chapter, we assume that the nodes have a class label, such as "liberal" or "conservative."

Network effects are very powerful, resulting even in popular proverbs such as "birds of a feather flock together" or "opposites attract." For example, in social networks, we often observe *homophily*: obese people tend to have obese friends [47], happy people tend to make their friends happy too [72], and in general, people tend to associate with like-minded friends, with respect to politics, hobbies, religion, etc. Homophily is encountered in other settings too: If a user likes some pages, she would probably like other pages that are heavily connected to her favorite ones (*personalized PageRank*); if a user likes some products, he will probably like similar products too (*content-based recommendation systems*); if a user is dishonest, his/her contacts are probably dishonest too (*accounting and calling-card fraud*). Occasionally, the reverse, called *heterophily*, is true. For example, in an online dating site, we may observe that talkative people prefer to date silent ones, and vice versa. Thus, by knowing the labels of a few nodes in a network, as well as whether homophily or heterophily applies in a given scenario, we can usually give good predictions about the labels of the remaining nodes.

In this chapter we start by covering the case with two classes (e.g., talkative vs. silent), and focus on finding the most likely class labels for all the nodes in the graph. Then, we extend our work to cover cases with more than two classes as well. Informally, the problem is defined as follows.

Problem 3.1 Guilt-by-association—Informal Given a graph with n nodes and m edges; n_+ and n_- nodes labeled as members of the positive and negative class, respectively. Find the class memberships of the rest of the nodes, assuming that neighbors influence each other.

The influence can be homophily or heterophily. This learning scenario, where we reason from observed training cases directly to test cases, is also called *transductive inference*, as opposed to *inductive learning*, where the training cases are used to infer general rules which are then applied to new test cases.

There are several, closely related methods that are used for transductive inference in networked data. Most of them handle homophily, and some of them also handle heterophily. We focus on three methods: Personalized PageRank (a.k.a. "Personalized Random Walk with Restarts," or just RWR) [93]; Semi-Supervised Learning (SSL) [235]; and Belief Propagation (BP) [172]. How are these methods related? Are they identical? If not, which method gives the best accuracy? Which method has the best scalability?

These questions are the focus of this chapter. We show that the three methods are related, but not identical, and present FaBP (or Fast Belief Propagation), a fast, accurate, and scalable algorithm with convergence guarantees.

3.1   GUILT-BY-ASSOCIATION TECHNIQUES

In this section, we provide background information for three guilt-by-association methods: RWR, SSL, and BP.

3.1.1   RANDOM WALK WITH RESTARTS (RWR)

RWR is the method underlying Google's classic PageRank algorithm [35], which is used to measure the relative importance of web pages. The idea behind PageRank is that there is an imaginary surfer who is randomly clicking on links. At any step, the surfer clicks on a link with probability c. The PageRank of a page is defined recursively and depends on the PageRank and the number of the web pages pointing to it; the more pages with high importance link to a page, the more important the page is. The PageRank vector $\mathbf{r}$ is defined to be the solution to the linear system:

$$\mathbf{r} = (1 - c)\mathbf{y} + c\mathbf{B}\mathbf{r},$$

where $1 - c$ is the restart probability and $c \in [0, 1]$. In the original algorithm, $\mathbf{B} = \mathbf{D}^{-1}\mathbf{A}$ which corresponds to the column-normalized adjacency matrix of the graph, and the starting vector is defined as $\mathbf{y} = \frac{1}{n}$ (uniform starting vector, where $\mathbf{1}$ is the all-ones column-vector).

A variation of PageRank is the *lazy random walks* [156] that allows the surfer to remain at the same page at any point. This option is encoded by the matrix $\mathbf{B} = \frac{1}{2}(\mathbf{I} + \mathbf{D}^{-1}\mathbf{A})$. The two methods are equivalent up to a change in c [14]. Another variation of PageRank includes works where the starting vector is not uniform, but depends on the topic distribution [93, 98]. These vectors are called *Personalized PageRank* vectors and they provide personalized or context-sensitive search ranking. Several works focus on speeding up RWR [71, 164, 209]. Related methods for node-to-node distance (but not necessarily *guilt-by-association*) include [122], parameterized by *escape probability* and *round-trip probability*, SimRank [98] and extensions/improvements [136, 230]. Although RWR is primarily used for scoring nodes relative to seed nodes, a formulation where RWR classifies nodes in a semi-supervised setting has been introduced by Lin and Cohen [140]. The connection between random walks and electric network theory is described in Doyle and Snell's book [57].

3.1.2   SEMI-SUPERVISED LEARNING (SSL)

SSL approaches are divided into four categories [235]: (i) *low-density separation* methods, (ii) *graph-based* methods, (iii) methods for *changing the representation*, and (iv) *co-training* methods. A survey of various SSL approaches is given in [235], and an application of transductive SSL for multi-label classification in heterogeneous information networks is described in [99]. SSL uses both labeled and unlabeled data for training, as opposed to supervised learning that uses only labeled data, and unsupervised learning that uses only unlabeled data. The principle behind SSL is that unlabeled data can help us decide the "metric" between data points and improve models' performance. SSL can be either transductive (it works only on the labeled and unlabeled *training* data) or inductive (it can be used to classify unseen data).

Most graph-based SSL methods are transductive, nonparametric, and discriminative. The graphs used in SSL consist of labeled and unlabeled nodes (examples), and the edges represent the similarity between them. SSL can be expressed in a regularization framework, where the goal is to estimate a function f on the graph with two parts.

- *Loss function*, which expresses that f is smooth on the whole graph (equivalently similar nodes are connected—"homophily").
- *Regularization*, which forces the final labels for the labeled examples to be close to their initial labels.

Here we refer to a variant of graph mincut introduced in [29]. Mincut is the mode of a Markov random field with binary labels (Boltzmann machine). Given l labeled points (x_i, y_i), $i = 1, \ldots, l$, and u unlabeled points $x_{l+1}, \ldots, x_{l+u}$, the final labels $\mathbf{x}$ are found by minimizing the energy function:

$$\alpha \sum_{j \in N(i)} a_{ij}(x_i - x_j)^2 + \sum_{1 \leq i \leq l} (y_i - x_i)^2, \tag{3.1}$$

where α is related to the coupling strength (homophily) of neighboring nodes, $N(i)$ denotes the neighbors of i, and a_{ij} is the $(i, j)^{th}$ element of the adjacency matrix $\mathbf{A}$.

3.1.3   BELIEF PROPAGATION (BP)

Belief Propagation (BP), also called the sum-product algorithm, is an *exact* inference method for graphical models with a tree structure [173]. In a nutshell, BP is an iterative message-passing algorithm that computes the marginal probability distribution for each unobserved node conditional on observed nodes: all nodes receive messages from their neighbors in parallel, update their belief states, and finally send new messages back out to their neighbors. In other words, at iteration t of the algorithm, the posterior belief of a node i is conditioned on the evidence of its t-step away neighbors in the underlying network. This process repeats until convergence and is well understood on trees.

When applied to loopy graphs, however, BP is not guaranteed to converge to the marginal probability distribution, nor to converge at all. In these cases it can be used as an approximation scheme [173], however. Despite the lack of exact convergence criteria, "loopy BP" has been shown to give accurate results *in practice* [228], and it is thus widely used today in various applications, such as error-correcting codes [130], stereo imaging in computer vision [66], fraud detection [150, 165], and malware detection [43]. Extensions of BP include *Generalized Belief Propagation* (GBP) that takes a multi-resolution viewpoint, grouping nodes into regions [229]; however, how to construct good regions is still an open research problem. Thus, we focus on standard BP, which is better understood.

We are interested in BP because not only is it an efficient inference algorithm on probabilistic graphical models, but it has also been successfully used for *transductive inference*. Our goal is to find the most likely beliefs (or classes) for all the nodes in a network. BP helps to iteratively propagate the information from a few nodes with initial (or explicit) beliefs throughout the network. More formally, consider a graph of n nodes and $k = 2$ possible class labels. The original update formulas [228] for the messages sent from node i to node j and the belief of node i for being in state x_i are

$$m_{ij}(x_j) \leftarrow \sum_{x_i} \phi_i(x_i) \cdot \psi_{ij}(x_i, x_j) \cdot \prod_{n \in N(i) \setminus j} m_{ni}(x_i) \tag{3.2}$$

$$b_i(x_i) \leftarrow \eta \cdot \phi_i(x_i) \cdot \prod_{j \in N(i)} m_{ji}(x_i), \tag{3.3}$$

where the message from node i to node j is computed based on all the messages sent by all its neighbors in the previous step except for the previous message sent from node j to node i. $N(i)$ denotes the neighbors of i, η is a normalization constant that guarantees that the beliefs sum to one, m_{ji} is the message sent from node j to node i, $\sum_i b_i(x_i) = 1$, ψ_{ij} represents the *edge potential* when node i is in state x_i and node j is in state x_j, and $\phi_i(x_i)$ is the prior belief of node i about being in state x_i. We will often refer to ψ_{ij} as homophily or coupling strength between nodes i and j. We can organize the edge potentials in a matrix form, which we call propagation or coupling matrix. We note that, by definition (Figure 3.1a), the propagation matrix is a right stochastic matrix (i.e., its rows add up to 1). Figure 3.1 illustrates two example propagation matrices that capture homophily and heterophily between two states. The entries in the propagation matrices are usually set based on expertise in a specific application domain, but there are also works that investigate methods for automatically learning the edge potentials from the data. For BP, the above update formulas are repeatedly computed for each node until the values (hopefully) converge to the final beliefs.

In this chapter, we study how the parameter choices for BP helps accelerate the algorithms, and how to implement the method on top of HADOOP [91] (open-source MapReduce implementation). This focus differentiates our work from existing research which speeds up BP by exploiting the graph structure [44, 165] or the order of message propagation [83].

Class of Receiver i

Class of sender j		+	−
	+	$P[x_i = +\|x_j = +]$	$P[x_i = -\|x_j = +]$
	−	$P[x_i = +\|x_j = -]$	$P[x_i = -\|x_j = -]$

(a) Explanation of Entries

	D	R
D	0.8	0.2
R	0.2	0.8

(b) Homophily

	T	S
T	0.3	0.7
S	0.7	0.3

(c) Heterophily

Figure 3.1: Propagation or coupling matrices capturing the network effects. (a) Explanation of the entries in the propagation matrix. P stands for probability; x_i and x_j represent the state/class/label of node i and j, respectively. Color intensity corresponds to the coupling strengths between classes of neighboring nodes. (b)–(c) Examples of propagation matrices capturing different network effects. (b) D: Democrats, R: Republicans. (c) T: Talkative, S: Silent.

3.1.4   SUMMARY

RWR, SSL, and BP have been used successfully in many tasks, such as ranking [35], classification [99, 235], malware and fraud detection [43, 150], and recommendation systems [114]. None of the above works, however, show the relationships between the three methods, or discuss their parameter choices (e.g., homophily scores). Table 3.1 qualitatively compares the three guilt-by-association methods and our proposed algorithm FABP: (i) all methods are scalable and support homophily; (ii) BP supports heterophily, but there is no guarantee on convergence; and (iii) our FABP algorithm improves on it to provide convergence. In the following discussion, we use the symbols that are defined in Table 3.2.

Table 3.1: Qualitative comparison of "guilt-by-association" (GBA) methods

GbA Method	Heterophily	Scalability	Convergence
RWR	✗	✓	✓
SSL	✗	✓	✓
BP	✓	✓	?
FABP	✓	✓	✓

3.2   FABP: FAST BELIEF PROPAGATION

In this section we present the three main formulas that show the similarity of the following methods: binary BP, and specifically our proposed approximation, linearized BP (FABP); Personalized Random Walk with Restarts (RWR); BP on Gaussian Random Fields (GABP); and Semi-Supervised Learning (SSL).

Table 3.2: FABP: Major symbols and definitions (matrices: in bold capital font; vectors in bold lower-case; scalars in plain font)

Symbols	Description
$m_{ij}, m(i, j)$	Message sent from node i to node j
$m_h, (i, j)$	$= m(i, j) - 0.5$, "about-half" message sent from node i to node j
ϕ	$n \times 1$ vector of the BP prior beliefs $\phi(i)\{>0.5; <0.5\}$ for $i \in \{\text{"+", "-"}\}$ class & $\phi(i) = 0$ means unknown class
ϕ_h	$= \phi - 0.5$, $n \times 1$ "about-half" prior belief vector
$\mathbf{b}$	$n \times 1$ BP final belief vector $b(i)\{>0.5; <0.5\}$ for $i \in \{\text{"+", "-"}\}$ class & $b(i) = 0$ means unclassified (neutral)
$\mathbf{b}_h$	$= \mathbf{b} - 0.5$, $n \times 1$ "about-half" final belief vector
h_h	$= h - 0.5$, "about-half" homophily factor, Where $h = \psi(\text{"+", "+"})$: entry of BP propagation matrix $h \to 0$ means strong heterophily and $h \to 1$ means strong homophily

For the homophily case, all the above methods are similar in spirit, and closely related to diffusion processes. Assuming that the positive class corresponds to green color and the negative class to red color, the n_+ nodes that belong to the positive class act as if they taint their neighbors with green color, and similarly do the negative nodes with red color. Depending on the strength of homophily, or equivalently the speed of diffusion of the color, eventually we have green-ish neighborhoods, red-ish neighborhoods, and bridge nodes (half-red, half-green).

As we show next, the solution vectors for each method obey very similar equations: they all involve solving a system of linear equations, where the matrix consists of a diagonal matrix plus a weighted or normalized version of the adjacency matrix. Table 3.3 shows the resulting equations, carefully aligned to highlight the correspondences.

Next we give the equivalence results for all three methods, and the convergence analysis for FABP. The convergence of Gaussian BP, a variant of Belief Propagation, is studied in [148] and [221]. The reasons that we focus on BP are that (a) it has a solid, Bayesian foundation, and (b) it is more general than the rest, being able to handle heterophily (as well as multiple-classes, on which we elaborate in Section 3.3).

Theorem 3.2 FABP. *The solution to belief propagation can be approximated by the linear system*

$$[\mathbf{I} + a\mathbf{D} - c'\mathbf{A}]\mathbf{b}_h = \phi_h \,, \tag{3.4}$$

Table 3.3: Main results, to illustrate correspondence. Matrices (in capital and bold) are $n \times n$; vectors (lower-case bold) are $n \times 1$ column vectors, and scalars (in lower-case plain font) typically correspond to strength of influence. Detailed definitions: in the text.

Method	Matrix	Unknown	Known
RWR	$[\mathbf{I} - c\mathbf{A}\mathbf{D}^{-1}] \times$	$\mathbf{x}$	$(1-c)\,\mathbf{y}$
SSL	$[\mathbf{I} + \alpha(\mathbf{D} - \mathbf{A}] \times$	$\mathbf{x}$	$\mathbf{y}$
GABP = SSL	$[\mathbf{I} + \alpha(\mathbf{D} - \mathbf{A}] \times$	$\mathbf{x}$	$\mathbf{y}$
FABP	$[\mathbf{I} + a(\mathbf{D} - c'\mathbf{A}] \times$	$\mathbf{b}_{\mathrm{h}}$	$\boldsymbol{\phi}_h$

where $a = 4h_h^2/(1 - 4h_h^2)$ and $c' = 2h_h/(1 - 4h_h^2)$; h_h is the "about-half" homophily factor which represents the notion of the propagation matrix; $\boldsymbol{\phi}_h$ is the vector of prior node beliefs, where $\boldsymbol{\phi}_h(i) = 0$ for the nodes with no explicit initial information; $\mathbf{b}_{\mathrm{h}}$ is the vector of final node beliefs.

Proof. The derivation of the FABP equation is given in Section 3.2.1.   □

Lemma 3.3   Personalized RWR   *The linear system for RWR given an observation* $\mathbf{y}$ *is described by the following formula:*

$$[\mathbf{I} - c\mathbf{A}\mathbf{D}^{-1}]\mathbf{x} = (1 - c)\mathbf{y}, \tag{3.5}$$

where $\mathbf{y}$ *is the starting vector and* $1 - c$ *is the restart probability,* $c \in [0, 1]$.

Proof. See [93], and [209].   □

   The starting vector $\mathbf{y}$ corresponds to the prior beliefs for each node in BP, with the small difference that $y_i = 0$ means that we know nothing about node i, while a positive score $y_i > 0$ means that the node belongs to the positive class (with the corresponding strength). In Section 5.1.2, we elaborate on the equivalence of RWR and FABP (Lemma 5.2, p. 99 and Theorem 5.3, p. 100). A connection between personalized PageRank and inference in tree-structured Markov random fields is drawn by Cohen [48].

Lemma 3.4   SSL   *Suppose we are given* l *labeled nodes* (x_i, y_i), $i = 1, \ldots, l$, $y_i \in \{0, 1\}$, *and* u *unlabeled nodes* $(x_{l+1}, \ldots, x_{l+u})$. *The solution to an SSL problem is given by the linear system:*

$$[\alpha(\mathbf{D} - \mathbf{A}) + \mathbf{I}]\mathbf{x} = \mathbf{y}, \tag{3.6}$$

where α *is related to the coupling strength (homophily) of neighboring nodes,* $\mathbf{y}$ *is the label vector for the labeled nodes, and* $\mathbf{x}$ *is the vector of final labels.*

Proof. Given l labeled points (x_i, y_i), $i = 1, \ldots, l$, and u unlabeled points $x_{l+1}, \ldots, x_{l+u}$ for a semi-supervised learning problem, based on an energy minimization formulation, we solve for the labels x_i by minimizing the following functional E

$$E(\mathbf{x}) = \alpha \sum_{j \in N(i)} a_{ij}(x_i - x_j)^2 + \sum_{1 \leq i \leq l} (y_i - x_i)^2, \tag{3.7}$$

where α is related to the coupling strength (homophily), of neighboring nodes. $N(i)$ denotes the neighbors of i. If *all* points are labeled, in matrix form, the functional can be re-written as

$$\begin{aligned} E(\mathbf{x}) &= \mathbf{x}^T[\mathbf{I} + \alpha(\mathbf{D} - \mathbf{A})]\mathbf{x} - 2\mathbf{x} \cdot \mathbf{y} + K(\mathbf{y}) \\ &= (\mathbf{x} - \mathbf{x}^*)^T[\mathbf{I} + \alpha(\mathbf{D} - \mathbf{A})](\mathbf{x} - \mathbf{x}^*) + K'(\mathbf{y}), \end{aligned}$$

where $\mathbf{x}^* = [\mathbf{I} + \alpha(\mathbf{D} - \mathbf{A})]^{-1}\mathbf{y}$, and K, K' are some constant terms which depend only on $\mathbf{y}$. Clearly, E achieves the minimum when

$$\mathbf{x} = \mathbf{x}^* = [\mathbf{I} + \alpha(\mathbf{D} - \mathbf{A})]^{-1}\mathbf{y}.$$

SSL is explained in Section 3.1.2 and in more depth in [235]. The connection of SSL and BP on Gaussian Random Fields (GABP) can be found in [235, 236]. □

As before, $\mathbf{y}$ represents the labels of the labeled nodes and, thus, it is related to the prior beliefs in BP; $\mathbf{x}$ corresponds to the labels of all the nodes or, equivalently, the final beliefs in BP.

Lemma 3.5   R-S correspondence   *On a regular graph (i.e., all nodes have the same degree d), RWR and SSL can produce identical results if*

$$\alpha = \frac{c}{(1-c)d}. \tag{3.8}$$

That is, we need to carefully align the homophily strengths α and c.

Proof. Based on Equations (3.5) and (3.6), the two methods will give identical results if

$$\begin{aligned} (1-c)[\mathbf{I} - c\mathbf{D}^{-1}\mathbf{A}]^{-1} &= [\mathbf{I} + \alpha(\mathbf{D} - \mathbf{A})]^{-1} \Leftrightarrow \\ (\frac{1}{(1-c)}\mathbf{I} - \frac{c}{(1-c)}\mathbf{D}^{-1}\mathbf{A})^{-1} &= (\alpha(\mathbf{D} - \mathbf{A}) + \mathbf{I})^{-1} \Leftrightarrow \\ \left(\frac{1}{1-c}\right)\mathbf{I} - \left(\frac{c}{1-c}\right)\mathbf{D}^{-1}\mathbf{A} &= \mathbf{I} + \alpha(\mathbf{D} - \mathbf{A}) \Leftrightarrow \\ \left(\frac{c}{1-c}\right)\mathbf{I} - \left(\frac{c}{1-c}\right)\mathbf{D}^{-1}\mathbf{A} &= \alpha(\mathbf{D} - \mathbf{A}) \Leftrightarrow \\ \left(\frac{c}{1-c}\right)[\mathbf{I} - \mathbf{D}^{-1}\mathbf{A}] &= \alpha(\mathbf{D} - \mathbf{A}) \Leftrightarrow \\ \left(\frac{c}{1-c}\right)\mathbf{D}^{-1}[\mathbf{D} - \mathbf{A}] &= \alpha(\mathbf{D} - \mathbf{A}) \Leftrightarrow \\ \left(\frac{c}{1-c}\right)\mathbf{D}^{-1} &= \alpha\mathbf{I}. \end{aligned}$$

If the graph is "regular," $d_i = d$ ($i = 1, \ldots$), or $\mathbf{D} = d \cdot \mathbf{I}$, in which case the condition becomes

$$\alpha = \frac{c}{(1-c)d} \Rightarrow c = \frac{\alpha d}{1 + \alpha d} \tag{3.9}$$

where d is the common degree of all the nodes. □

Although Lemma 3.5 refers to regular graphs, the result can be extended to arbitrary graphs as well. In this case, instead of having a single homophily strength α, we introduce a homophily factor per node i, $\alpha_i = \frac{c}{(1-c)d_i}$ with degree d_i. The connection between RWR and SSL is further explained in [79].

ARITHMETIC EXAMPLES

In this section we illustrate that SSL and RWR give closely related solutions. We set α to be $\alpha = \frac{c}{(1-c)*\bar{d}}$ (where $\bar{d}$ is the average degree).

Figure 3.2 shows the scatter plot: each red star (x_i, y_i) corresponds to a node, say, node i; the coordinates are the RWR and SSL scores, respectively. The blue circles correspond to the perfect identity, and thus are on the 45° line. The left plot in Figure 3.2 has three major groups, corresponding to the "+"-labeled, the unlabeled, and the "-"-labeled nodes (from top-right to bottom-left, respectively). The right plot in Figure 3.2 shows a magnification of the central part (the unlabeled nodes). Notice that the red stars are close to the 45° line. The conclusion is that (a) the SSL and RWR scores are similar, and (b) the rankings are the same: whichever node is labeled as "positive" by SSL, gets a high score by RWR, and conversely.

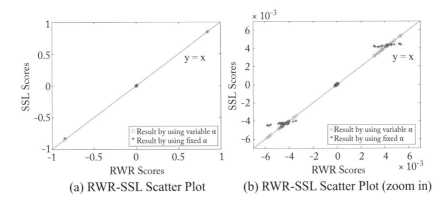

(a) RWR-SSL Scatter Plot   (b) RWR-SSL Scatter Plot (zoom in)

Figure 3.2: Illustration of near-equivalence of SSL and RWR. We show the SSL scores vs. the RWR scores for the nodes of a random graph; blue circles (ideal, perfect equality) and red stars (real). Right: a zoom-in of the left. Most red stars are on or close to the diagonal: the two methods give similar scores, and identical assignments to positive/negative classes.

3.2.1   DERIVATION

We now derive FABP, our closed formula for approximating belief propagation (Theorem 3.2). The key ideas are to center the values around $\frac{1}{2}$, allow only small deviations, and use, for each variable, the *odds* of the positive class.

The propagation or coupling matrix (see Section 3.1.3 and Figure 3.1) is central to BP as it captures the network effects—i.e., the edge potential (or strength) between states/classes/labels. Often the propagation matrix is symmetric, that is, the edge potential does not depend on the direction in which a message is transmitted (e.g., Figure 3.1c). Moreover, because the propagation matrix is also left stochastic, we can entirely describe it with a single value, such as the first element $P[x_i = +|x_j = +] = \psi(\text{"+"}, \text{"+"})$. We denote this value with h_h and call it "about-half" homophily factor (Table 3.2).

In more detail, the key ideas for the upcoming proofs are as follows.

1. We center the values of all the quantities involved in BP around $\frac{1}{2}$. Specifically, we use the vectors $\mathbf{m_h} = \mathbf{m} - \frac{1}{2}$, $\mathbf{b_h} = \mathbf{b} - \frac{1}{2}$, $\boldsymbol{\phi_h} = \boldsymbol{\phi} - \frac{1}{2}$, and the scalar $h_h = h - \frac{1}{2}$. We allow the values to deviate from the $\frac{1}{2}$ point by a small constant ϵ, use the MacLaurin expansions in Table 3.4, and keep only the first order terms. By doing so, we avoid the sigmoid/non-linear equations of BP.

2. For each quantity p, we use the *odds* of the positive class, $p_r = p/(1-p)$, instead of probabilities. This results in only one value for node i, $p_r(i) = \frac{p_+(i)}{p_-(i)}$ instead of two. Moreover, the normalization factor is no longer needed.

Table 3.4: Logarithm and division approximations used in the derivation of FABP

	Formula	Maclaurin Series	Approximation
Logarithm	$\ln(1 + \epsilon)$	$= \epsilon - \frac{\epsilon^2}{2} + \frac{\epsilon^3}{3} - \ldots$	$\approx \epsilon$
Division	$\frac{1}{1-\epsilon}$	$= 1 + \epsilon + \epsilon^2 + \epsilon^3 + \ldots$	$\approx 1 + \epsilon$

We start from the original BP equations given in Section 3.1.3, and apply our two main ideas to obtain the odds expressions for the BP message and belief equations. In the following equations, we use the notation $var(i)$ to denote that the variable refers to node i. We note that in the original BP equations (Equations (3.2) and (3.3)), we had to write $var_i(x_i)$ to denote that the var refers to node i and the state x_i. However, by introducing the odds of the positive class, we no longer need to note the state/class of node i.

Lemma 3.6   *Expressed as ratios, the BP equations for the message and beliefs become:*

$$m_r(i, j) \leftarrow B[h_r, b_{r,adjusted}(i, j)] \tag{3.10}$$

$$b_r(i) \leftarrow \phi_r(i) \cdot \prod_{j \in N(i)} m_r(j, i), \tag{3.11}$$

where $b_{r,adjusted}(i, j) = b_r(i)/m_r(j, i)$, and $B(x, y) = \frac{x \cdot y + 1}{x + y}$ is a blending function.

Proof. Based on the notations introduced in our second key idea, $b_+(i) = b_i(x_i = +)$ in Equation (3.3). By writing out the definition of b_r and using Equation (3.3) in the numerator and denominator, we obtain

$$b_r(i) = \frac{b_+(i)}{b_-(i)} \overset{Equation\ (3.3)}{=} \frac{\eta \cdot \phi_+(i) \cdot \prod_{j \in N(i)} m_+(j, i)}{\eta \cdot \phi_-(i) \cdot \prod_{j \in N(i)} m_-(j, i)}$$

$$= \phi_r(i) \prod_{j \in N(i)} \frac{m_+(j, i)}{m_-(j, i)}$$

$$= \phi_r(i) \prod_{j \in N(i)} m_r(j, i).$$

This concludes the proof for Equation (3.11). We can write Equation (3.11) in the following form, which we will use later to prove Equation (3.10):

$$b_r(i) = \phi_r(i) \prod_{j \in N(i)} m_r(j, i) \Rightarrow \prod_{n \in N(i) \backslash j} m_r(n, i) m_r(j, i) = \frac{b_r(i)}{\phi_r(i)} \Rightarrow$$

$$\prod_{n \in N(i) \backslash j} m_r(n, i) = \frac{b_r(i)}{\phi_r(i) m_r(j, i)}. \tag{3.12}$$

To obtain Equation (3.10), we start by writing out the definition of $m_r(i, j)$ and then substitute Equation (3.2) in the numerator and denominator by considering in the $\sum_{x_i}$ all possible states $x_i = \{+, -\}$:

$$m_r(i, j) = \frac{m_+(i, j)}{m_-(i, j)}$$

$$\overset{Equation\ (3.2)}{=} \frac{\sum_{x=\{+,-\}} \phi_x(i) \cdot \psi_{ij}(x, +) \cdot \prod_{n \in N(i) \backslash j} m_x(n, i)}{\sum_{x=\{+,-\}} \phi_x(i) \cdot \psi_{ij}(x, -) \cdot \prod_{n \in N(i) \backslash j} m_x(n, i)}$$

$$= \frac{\phi_+(i) \cdot \psi_{ij}(+, +) \cdot \prod_{n \in N(i) \backslash j} m_+(n, i) + \phi_-(i) \cdot \psi_{ij}(-, +) \cdot \prod_{n \in N(i) \backslash j} m_-(n, i)}{\phi_+(i) \cdot \psi_{ij}(+, -) \cdot \prod_{n \in N(i) \backslash j} m_+(n, i) + \phi_-(i) \cdot \psi_{ij}(-, -) \cdot \prod_{n \in N(i) \backslash j} m_-(n, i)}.$$

From the definition of h in Table 3.2 and our second main idea, we get that $h_+ = \psi(+, +) = \psi(-, -)$ (which is independent of the nodes, as we explained at the beginning of this section), while $h_- = \psi(+, -) = \psi(-, +)$. By substituting these formulas in the last equation and by dividing with $d_{aux}(i) = \phi_+(i)h_- \prod_{n \in N(i) \setminus j} m_-(n,i)$, we get

$$m_r(i,j) = \frac{\phi_+(i) \cdot h_+ \cdot \prod_{n \in N(i) \setminus j} m_+(n,i) + \phi_-(i) \cdot h_- \cdot \prod_{n \in N(i) \setminus j} m_-(n,i)}{\phi_+(i) \cdot h_- \cdot \prod_{n \in N(i) \setminus j} m_+(n,i) + \phi_-(i) \cdot h_+ \cdot \prod_{n \in N(i) \setminus j} m_-(n,i)}$$

$$\underset{=}{\div d_{aux}(i)} \frac{h_r + \frac{1}{\phi_r(i) \prod_{n \in N(i) \setminus j} m_r(n,i)}}{1 + \frac{h_r}{\phi_r(i) \prod_{n \in N(i) \setminus j} m_r(n,i)}}$$

$$\overset{Equation\ (3.12)}{=} \frac{h_r + \frac{m_r(j,i)}{b_r(i)}}{1 + \frac{h_r m_r(j,i)}{b_r(i,j)}} = \frac{h_r \frac{b_r(i)}{m_r(j,i)} + 1}{h_r + \frac{b_r(i)}{m_r(j,i)}} = \frac{h_r b_{r,adjusted}(i,j) + 1}{h_r + b_{r,adjusted}(i,j)} = B[h_r, b_{r,adjusted}(i,j)].$$

We note that the division by $m_r(j,i)$ (which comes from Equation (3.12)) subtracts the influence of node j when preparing the message $m(i,j)$. The same effect is captured by the original message equation (Equation (3.2)). □

Before deriving the "about-half" beliefs $\mathbf{b_h}$ and "about-half" messages $\mathbf{m_h}$, we introduce some approximations for all the quantities of interest (messages and beliefs) that will be useful in the remaining proofs.

Lemma 3.7 Approximations *Let $\{v_r, a_r, b_r\}$ and $\{v_h, a_h, b_h\}$ be the odds and "about-half" values of the variables v, a, and b, respectively. The following approximations are fundamental for the rest of our lemmas, and hold for all the variables of interest (m_r, b_r, ϕ_r, and h_r).*

$$v_r = \frac{v}{1-v} = \frac{1/2 + v_h}{1/2 - v_h} \approx 1 + 4v_h \tag{3.13}$$

$$B(a_r, b_r) \approx 1 + 8a_h b_h, \tag{3.14}$$

where $B(a_r, b_r) = \frac{a_r \cdot b_r + 1}{a_r + b_r}$ is the blending function for any variables a_r, b_r.

Sketch of proof. For Equation (3.13), we start from the odds and "about-half" approximations ($v_r = \frac{v}{1-v}$ and $v_h = v - \frac{1}{2}$ resp.), and then use the Maclaurin series expansion for division (Table 3.4) and keep only the first order terms:

$$v_r = \frac{v}{1-v} = \frac{1/2 + v_h}{1/2 - v_h}$$

$$= \frac{1 + 2v_h}{1 - 2v_h} \overset{Table\ 3.4}{\approx} (1 + 2v_h)(1 + 2v_h)$$

$$= 1 + 4v_h^2 + 4v_h \overset{1^{st}\ order}{=} 1 + 4v_h \Rightarrow$$

$$v_r \approx 1 + 4v_h.$$

For Equation (3.14), we start from the definition of the blending function, then apply Equation (3.13) for all the variables, and use the Maclaurin series expansion for division (Table 3.4). As before, we keep only the first-order terms in our approximation:

$$B(a_r, b_r) = \frac{a_r \cdot b_r + 1}{a_r + b_r} \overset{Equation\ (3.13)}{=} \frac{(1 + 4a_h)(1 + 4b_h) + 1}{(1 + 4a_h) + (1 + 4b_h)}$$

$$= 1 + \frac{16a_h b_h}{2 + 4a_h + 4b_h} = 1 + \frac{8a_h b_h}{1 + 2(a_h + b_h)}$$

$$\overset{Table\ 3.4}{\approx} 1 + 8a_h b_h (1 - 2(a_h + b_h)) = 1 + 8a_h b_h \Rightarrow$$

$$B(a_r, b_r) \approx 1 + 8a_h b_h.$$

□

The following three lemmas are useful in order to derive the linear equation of FABP. We note that in all the lemmas we apply several approximations in order to linearize the equations; we omit the "≈" symbol so that the proofs are more readable.

Lemma 3.8   *The about-half version of the belief equation becomes, for small deviations from the half-point:*

$$b_h(i) \approx \phi_h(i) + \sum_{j \in N(i)} m_h(j, i). \tag{3.15}$$

Proof. We use Equation (3.11) and Equation (3.13), and apply the appropriate MacLaurin series expansions:

$$b_r(i) = \phi_r(i) \prod_{j \in N(i)} m_r(j, i) \Rightarrow$$

$$log\,(1 + 4b_h(i)) = log\,(1 + 4\phi_h(i)) + \sum_{j \in N(i)} log\,(1 + 4m_h(j, i)) \Rightarrow$$

$$b_h(i) = \phi_h(i) + \sum_{j \in N(i)} m_h(j, i).$$

□

Lemma 3.9   *The about-half version of the message equation becomes:*

$$m_h(i, j) \approx 2h_h[b_h(i) - m_h(j, i)]. \tag{3.16}$$

Proof. We combine Equation (3.10), Equation (3.13), and Equation (3.14):

$$m_r(i, j) = B[h_r, b_{r,adjusted}(i, j)] \Rightarrow m_h(i, j) = 2h_h b_{h,adjusted}(i, j). \qquad (3.17)$$

In order to derive $b_{h,adjusted}(i, j)$ we use Equation (3.13) and the approximation of the MacLaurin expansion $\frac{1}{1+\epsilon} = 1 - \epsilon$ for a small quantity ϵ:

$$
\begin{aligned}
b_{r,adjusted}(i, j) &= b_r(i)/m_r(j, i) \Rightarrow \\
1 + b_{h,adjusted}(i, j) &= (1 + 4b_h(i))(1 - 4m_h(j, i)) \Rightarrow \\
b_{h,adjusted}(i, j) &= b_h(i) - m_h(j, i) - 4b_h(i)m_h(j, i). \qquad (3.18)
\end{aligned}
$$

Substituting Equation (3.18) to Equation (3.17) and ignoring the terms of second order leads to the about-half version of the message equation. □

Lemma 3.10   *At steady state, the messages can be expressed in terms of the beliefs:*

$$m_h(i, j) \approx \frac{2h_h}{(1 - 4h_h^2)} [b_h(i) - 2h_h b_h(j)]. \qquad (3.19)$$

Proof. We apply Lemma 3.9 both for $m_h(i, j)$ and $m_h(j, i)$ and we solve for $m_h(i, j)$. □

Based on the above derivations, we can now obtain the equation for FABP (Theorem 3.2), which we presented in Section 3.2.

Proof. [for Theorem 3.2: FABP] By combining Equation (3.15) and Equation (3.19), we obtain:

$$b_h(i) - \sum_{j \in N(i)} m_h(j, i) = \phi_h(i) \Rightarrow$$

$$b_h(i) + \sum_{j \in N(i)} \frac{4h_h^2 b_h(j)}{1 - 4h_h^2} - \sum_{j \in N(i)} \frac{2h_h}{1 - 4h_h^2} b_h(i) = \phi_h(i) \Rightarrow$$

$$b_h(i) + a \sum_{j \in N(i)} b_h(i) - c' \sum_{j \in N(i)} b_h(j) = \phi_h(i) \Rightarrow$$

$$(\mathbf{I} + a\mathbf{D} - c'\mathbf{A})\mathbf{b_h} = \boldsymbol{\phi_h}.$$

□

3.2.2 ANALYSIS OF CONVERGENCE

Here we study the sufficient, but not necessary conditions for which our method, FABP, converges. The implementation details of FABP are described in the upcoming Section 3.2.3. Lemma 3.11, Lemma 3.12, and Equation (3.23) give the convergence conditions.

All our results are based on the power expansion that results from the inversion of a matrix of the form $\mathbf{I} - \mathbf{W}$; all the methods undergo this process, as we show in Table 3.3. Specifically, we need the inverse of the matrix $\mathbf{I} + a\mathbf{D} - c'\mathbf{A} = \mathbf{I} - \mathbf{W}$, which is given the expansion:

$$(\mathbf{I} - \mathbf{W})^{-1} = \mathbf{I} + \mathbf{W} + \mathbf{W}^2 + \mathbf{W}^3 + \cdots \tag{3.20}$$

and the solution of the linear system is given by the formula

$$(\mathbf{I} - \mathbf{W})^{-1}\boldsymbol{\phi_h} = \boldsymbol{\phi_h} + \mathbf{W} \cdot \boldsymbol{\phi_h} + \mathbf{W} \cdot (\mathbf{W} \cdot \boldsymbol{\phi_h}) + \cdots . \tag{3.21}$$

This method is fast, since the computation can be done in iterations, each one of which consists of a sparse-matrix/vector multiplication. This is referred to as the *Power Method*. However, the power method does not always converge. In this section we examine its convergence conditions.

Lemma 3.11   Largest eigenvalue   *The series $\sum\limits_{k=0}^{\infty} |\mathbf{W}|^k = \sum\limits_{k=0}^{\infty} |c'\mathbf{A} - a\mathbf{D}|^k$ converges iff $\lambda(\mathbf{W}) < 1$, where $\lambda(\mathbf{W})$ is the magnitude of the largest eigenvalue of $\mathbf{W}$.*

Given that the computation of the largest eigenvalue is non-trivial, we suggest using Lemma 3.12 or Lemma 3.13, which give a closed form for computing the "about-half" homophily factor, h_h.

Lemma 3.12   1-norm   *The series $\sum\limits_{k=0}^{\infty} |\mathbf{W}|^k = \sum\limits_{k=0}^{\infty} |c'\mathbf{A} - a\mathbf{D}|^k$ converges if*

$$h_h < \frac{1}{2(1 + \max_j (d_{jj}))} \tag{3.22}$$

where d_{jj} are the elements of the diagonal matrix D.

Proof. In a nutshell, the proof is based on the fact that the power series converges if the 1-norm, or equivalently the ∞-norm, of the symmetric matrix $\mathbf{W}$ is smaller than 1.

Specifically, in order for the power series to converge, a sub-multiplicative norm of matrix $\mathbf{W} = c\mathbf{A} - a\mathbf{D}$ should be smaller than 1. In this analysis we use the 1-norm (or equivalently the

∞-norm). The elements of matrix $\mathbf{W}$ are either $c = \frac{2h_h}{1-4h_h^2}$ or $-ad_{ii} = \frac{-4h_h^2 d_{ii}}{1-4h_h^2}$. Thus, we require

$$\max_j \left(\sum_{i=1}^n |A_{ij}| \right) < 1 \Rightarrow (c + a) \cdot \max_j d_{jj} < 1$$

$$\frac{2h}{1 - 2h} \max_j d_{jj} < 1 \Rightarrow h_h < \frac{1}{2(1 + \max_j d_{jj})}.$$

$\square$

Lemma 3.13 Frobenius norm *The series* $\sum_{k=0}^{\infty} |\mathbf{W}|^k = \sum_{k=0}^{\infty} |c'\mathbf{A} - a\mathbf{D}|^k$ *converges if*

$$h_h < \sqrt{\frac{-c_1 + \sqrt{c_1^2 + 4c_2}}{8c_2}} \qquad (3.23)$$

where $c_1 = 2 + \sum_i d_{ii}$ *and* $c_2 = \sum_i d_{ii}^2 - 1$.

Proof. This upper bound for h_h is obtained by considering the Frobenius norm of matrix $\mathbf{W}$ and solving the inequality $\| \mathbf{W} \|_F = \sqrt{\sum_{i=1}^n \sum_{j=1}^n |\mathbf{W}_{ij}|^2} < 1$ with respect to h_h. $\square$

Equation (3.23) is preferable over Equation (3.22) when the degrees of the graph's nodes demonstrate considerable standard deviation. The 1-norm yields small h_h for very big values of the highest degree, while the Frobenius norm gives a higher upper bound for h_h. Nevertheless, we should bear in mind that h_h should be a sufficiently small number in order for the "about-half" approximations to hold.

3.2.3 ALGORITHM

Based on the analysis in Section 3.2 and Section 3.2.2, we propose the FABP algorithm (Algorithm 3.1).

We conjecture that if the achieved accuracy is not sufficient, the results of FABP can still be a good starting point for the original, iterative BP algorithm. One would need to use the final beliefs of FABP as the prior beliefs of BP, and run a few iterations of BP until convergence. In the datasets we studied, this additional step was not required, as FABP achieves equal or higher accuracy than BP, while being faster.

Algorithm 3.1 FABP

Input : graph G, prior beliefs ϕ
Output : beliefs **b** for all the nodes

1: **Step 1:** Pick h_h to achieve convergence: $h_h = \max\{Equation\ (3.22), Equation\ (3.23)\}$ and compute the parameters a and c', as described in Theorem 3.2.
2: **Step 2:** Solve the linear system of Equation (3.4). Notice that all the quantities involved in this equation are close to zero.

3.3 EXTENSION TO MULTIPLE CLASSES

Now we briefly present an extension of FABP to $k \geq 2$ classes, while also capturing a mix of homophily and heterophily [78]. We illustrate with an example taken from online auction settings like eBay [165]: We observe $k=3$ classes of people: fraudsters (F), accomplices (A), and honest people (H). Honest people buy and sell from other honest people, as well as accomplices. Accomplices establish a good reputation (thanks to multiple interactions with honest people), they never interact with other accomplices (waste of effort and money), but they do interact with fraudsters, forming near-bipartite cores between the two classes. Fraudsters primarily interact with accomplices (to build reputation); the interaction with honest people (to defraud them) happens in the last few days before the fraudster's account is shut down.

Thus, in the general case, we can have k different classes, and $k \times k$ affinities or coupling strengths between them. These affinities can be organized in a coupling matrix, which we call *propagation matrix*. In Figure 3.3, we give a coupling matrix for the eBay scenario, which shows a more general example: We see homophily between members of class H and heterophily between members of classes A and F.

	H	A	F
H	0.6	0.3	0.1
A	0.3	0.0	0.7
F	0.1	0.7	0.2

Figure 3.3: General case of coupling matrix with mix of *network effects*. Color intensity corresponds to the coupling strengths between classes of neighboring nodes. H: Honest, A: Accomplice, F: Fraudster.

As in the case of two classes, here too we are interested in the most likely "beliefs" (or labels) for all nodes in the graph by using BP. The underlying problem is then: How can we assign class labels when we know who-contacts-whom and the *a priori* ("initial" or "explicit") labels for some of the nodes in the network? How can we handle multiple class labels, as well as intricate network effects?

We use the symbol $\mathbf{H}$ to represent the $k \times k$ matrix with the coupling weights (in the $k = 2$ case, we used h for the constant homophily factor). Concretely, if $H(i, i) > H(j, i)$ for all $i \neq j$, we say homophily is present. If the opposite inequality is true for all i, then heterophily is present. Otherwise, there exists homophily for some classes, and heterophily for others. Similarly to our previous analysis, we assume that the relative coupling between classes is the same in the whole graph; i.e., $H(j, i)$ is identical for all edges in the graph. We further require this coupling matrix $\mathbf{H}$ to be *doubly stochastic and symmetric*: (i) Double stochasticity is a necessary requirement for our mathematical derivation.[1] (ii) Symmetry is not required, but follows from our assumption of undirected edges.

The problem then becomes as follows.

Problem 3.14 Top belief assignment

- Given (1) an undirected graph with n nodes and adjacency matrix $\mathbf{A}$,
  (2) a symmetric, doubly stochastic coupling $k \times k$ matrix $\mathbf{H}$, where $H(j, i)$ indicates the relative influence of class j of a node on class i of its neighbor, and
  (3) a matrix of explicit beliefs $\mathbf{\Phi}$ ($\Phi(s, i) \neq 0$) is the belief in class i by node s.

- Find for *each* node a set of classes with the highest final belief (i.e., top belief assignment).

As in the specific case of $k = 2$, the main idea is to *center* the values around default values (using Maclaurin series expansions) and to then restrict our parameters to small deviations from these defaults. The resulting equations replace multiplication with addition and can thus be put into a matrix framework with a closed-form solution. Next, we give two definitions that are needed for our first main result.

Definition 3.15 Centering.   We call a vector or matrix $\mathbf{x}$ "*centered around c*" if all its entries are close to c and their average is exactly c.

Definition 3.16 Residual vector/matrix.   If a vector $\mathbf{x}$ is centered around c, then the residual vector around c is defined as $\hat{\mathbf{x}} = [x_1 - c, x_2 - c, \ldots]$. Accordingly, we denote a matrix $\hat{\mathbf{X}}$ as a residual matrix if each column and row vector corresponds to a residual vector.

---

[1]We note that single-stochasticity could easily be constructed by taking any set of vectors of relative coupling strengths between neighboring classes, normalizing them to 1, and arranging them in a matrix.

For example, we call the vector $\mathbf{x} = [1.01, 1.02, 0.97]$ centered around $c = 1$.[2] The residuals from c will form the *residual vector* $\hat{\mathbf{x}} = [0.01, 0.02, -0.03]$. Notice that the entries in a residual vector always sum up to 0, by construction.

The main result, which extends FABP to multiple classes, is as follows.

Theorem 3.17   Linearized BP (LinBP).   *Let $\mathbf{B}_h$ and $\mathbf{\Phi}_h$ be the residual matrices of final and explicit beliefs centered around $1/k$, $\mathbf{H}_h$ the residual coupling matrix centered around $1/k$, $\mathbf{A}$ the adjacency matrix, and $\mathbf{D} = diag(\mathbf{d})$ the diagonal degree matrix. Then, the final belief assignment from belief propagation is approximated by the equation system:*

$$\mathbf{B}_h = \mathbf{\Phi}_h + \mathbf{A}\mathbf{B}_h\mathbf{H}_h - \mathbf{D}\mathbf{B}_h\mathbf{H}_h{}^2 \qquad (LinBP). \qquad (3.24)$$

In practice, we will solve Equation (3.24) via an iterative computation. However, this equation has a *closed-form* solution, which allows us to study the convergence of the iterative updates, and give the exact convergence criteria based on problem parameters. We need to introduce two new notions: Let $\mathbf{X}$ and $\mathbf{Y}$ be matrices of order $m \times n$ and $p \times q$, respectively, and let $\mathbf{x}_j$ denote the j-th column of matrix $\mathbf{X}$, i.e., $\mathbf{X} = \{x_{ij}\} = [\mathbf{x}_1 \ldots \mathbf{x}_n]$. First, the *vectorization* of matrix $\mathbf{X}$ stacks the columns of a matrix one underneath the other to form a single column vector:

$$\text{vec}(\mathbf{X}) = [\mathbf{x}_1, \ldots, \mathbf{x}_n]^T.$$

Second, the *Kronecker product* of $\mathbf{X}$ and $\mathbf{Y}$ is the $mp \times nq$ matrix defined by

$$\mathbf{X} \otimes \mathbf{Y} = \begin{bmatrix} x_{11}\mathbf{Y} & x_{12}\mathbf{Y} & \ldots & x_{1n}\mathbf{Y} \\ x_{21}\mathbf{Y} & x_{22}\mathbf{Y} & \ldots & x_{2n}\mathbf{Y} \\ \vdots & \vdots & \ddots & \vdots \\ x_{m1}\mathbf{Y} & x_{m2}\mathbf{Y} & \ldots & x_{mn}\mathbf{Y} \end{bmatrix}.$$

Based on these notations, we obtain the closed-form of LINBP in lemma 3.18.

Lemma 3.18   Closed-form LinBP   *The closed-form solution for LinBP (Equation (3.24)) is:*

$$\text{vec}(\mathbf{B}_h) = (\mathbf{I}_{nk} - \mathbf{H}_h \otimes \mathbf{A} + \mathbf{H}_h{}^2 \otimes \mathbf{D})^{-1}\text{vec}(\mathbf{\Phi}_h) \quad (LinBP). \qquad (3.25)$$

For details of derivations and in-depth analysis of LINBP and its convergence criteria, we refer the interested reader to [78]. Extensions of belief propation to heterogeneous graphs are presented in [62].

---

[2]All vectors $\mathbf{x}$ in this chapter are assumed to be *column vectors* $[x_1, x_2, \ldots]^T$ even if written as row vectors $[x_1, x_2, \ldots]$.

3.4   EMPIRICAL RESULTS

We present experimental results to answer the following questions:

Q1: How accurate is FABP?
Q2: Under what conditions does FABP converge?
Q3: How robust is FABP to the values of h and ϕ?
Q4: How does FABP scale on very large graphs with billions of nodes and edges?

The graphs we used in our experiments are summarized in Table 3.5. To answer Q1 (accuracy), Q2 (convergence), and Q3 (robustness), we use the DBLP dataset[3] [76],which consists of 14,376 papers, 14,475 authors, 20 conferences, and 8,920 terms. A small portion of these nodes are manually labeled based on their area (Artificial Intelligence, Databases, Data Mining and Information Retrieval): 4,057 authors, 100 papers, and all the conferences. We adapted the labels of the nodes to two classes: AI (Artificial Intelligence) and *not* AI (= Databases, Data Mining and Information Retrieval). In each trial, we run FABP on the DBLP network where $(1 - p)\% = (1 - a)\%$ of the labels of the papers and the authors have been discarded. Then, we test the classification accuracy on the nodes whose labels were discarded. The values of a and p are $0.1\%, 0.2\%, 0.3\%, 0.4\%, 0.5\%$, and 5%. To avoid combinatorial explosion, we consider $\{h_h, priors\} = \{\pm0.002, \pm0.001\}$ as the anchor values, and then we vary one parameter at a time. When the results are the same for different values of $a\% = p\%$, due to lack of space, we randomly pick the plots to present.

Table 3.5: FABP: Order and size of graphs

Dataset	Nodes	Edges
Yahoo Web	1,413,511,390	6,636,600,779
Kronecker 1	177,147	1,977,149,596
Kronecker 2	120,552	1,145,744,786
Kronecker 3	59,049	282,416,200
Kronecker 4	19,683	40,333,924
DBLP	37,791	170,794

To answer Q4 (scalability), we use the real YahooWeb graph and synthetic Kronecker graph datasets. YahooWeb is a Web graph containing 1.4 billion web pages and 6.6 billion edges; we label 11 million educational and 11 million adult web pages. We use 90% of these labeled data to set node priors, and use the remaining 10% to evaluate the accuracy. For parameters, we set h_h to 0.001 using Lemma 3.13 (Frobenius norm), and the magnitude of the prior beliefs to 0.5 ± 0.001. The Kronecker graphs are synthetic graphs generated by the Kronecker generator [131].

[3]http://web.engr.illinois.edu/~mingji1/DBLP_four_area.zip

3.4.1   ACCURACY

Figure 3.4 shows the scatter plots of beliefs (FABP vs. BP) for each node of the DBLP data. We observe that FABP and BP result in practically the same beliefs for all the nodes in the graph, when run with the same parameters, and thus, they yield the same accuracy. Conclusions are identical for any labeled-set-size we tried (0.1% and 0.3% shown in Figure 3.4).

Observation 3.19   FABP and BP agree on the classification of the nodes when run with the same parameters.

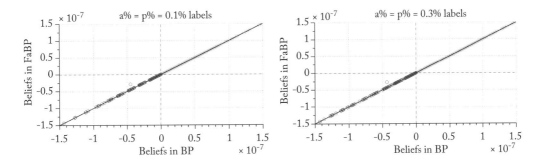

Figure 3.4: Scatter plot of beliefs for $(h, priors) = (0.5 + / - 0.0020, 0.5 + / - 0.001)$. The quality of scores of FABP is near-identical to BP, i.e., all on the 45° line in the scatter plot of beliefs (FABP vs. BP) for each node of the DBLP sub-network; red/green points correspond to nodes classified as "AI/not-AI," respectively.

3.4.2   CONVERGENCE

We examine how the value of the "about-half" homophily factor affects the convergence of FABP. Figure 3.5 the red line annotated with "max $|eval| = 1$" splits the plots into two regions: (a) on the left, the Power Method converges and FABP is accurate, (b) on the right, the Power Method diverges resulting in a significant drop in the classification accuracy. We annotate the number of classified nodes for the values of h_h that leave some nodes unclassified because of numerical representation issues. The low accuracy scores for the smallest values of h_h are due to the unclassified nodes, which are counted as misclassifications. The Frobenius norm-based method yields greater upper bound for h_h than the 1-norm-based method, preventing any numerical representation problems.

Observation 3.20   Our convergence bounds consistently coincide with high-accuracy regions. Thus, we recommend choosing the homophily factor based on the Frobenius norm using Equation (3.23).

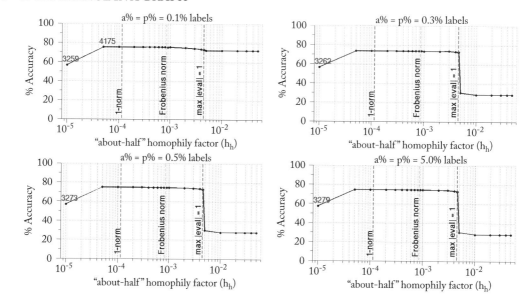

Figure 3.5: Accuracy with respect to h_h (*prior beliefs* $= \pm 0.00100$). FABP achieves maximum accuracy within the convergence bounds. If not all nodes are classified by FABP, we give the number of classified nodes in red.

3.4.3   ROBUSTNESS

Figure 3.5 shows that FABP is robust to the "about-half" homophily factor, h_h, as long as the latter is within the convergence bounds. In Figure 3.6 we observe that the accuracy score is insensitive to the *magnitude* of the prior beliefs. We only show the cases $a, p \in \{0.1\%, 0.3\%, 0.5\%\}$, as for all values except for $a, p = 5.0\%$, the accuracy is practically identical. The results are similar for different "about-half" homophily factors.

Observation 3.21   The accuracy results are insensitive to the *magnitude* of the prior beliefs and homophily factor—as long as the latter is within the convergence bounds we gave in Section 3.2.2.

3.4.4   SCALABILITY

To show the scalability of FABP, we implemented FABP on HADOOP, an open source MAPRE-DUCE framework which has been successfully used for large-scale graph analysis [106]. We first show the scalability of FABP on the number of edges of Kronecker graphs. As seen in Figure 3.7, FABP scales linear on the number of edges. Next, we compare the HADOOP implementation of

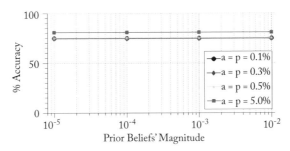

Figure 3.6: Accuracy with respect to the magnitude of the prior beliefs $h_h = \{\pm 0.00200)$. FABP is robust to the *magnitude* of the prior beliefs.

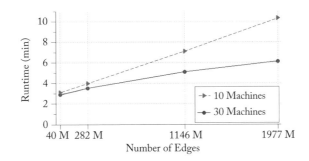

Figure 3.7: Running time of FABP vs. # edges for 10 and 30 machines on HADOOP. Kronecker graphs are used.

FABP and BP [102] in terms of running time and accuracy on the YahooWeb graph. Figure 3.8a–b shows that FABP achieves the maximum accuracy after two iterations of the Power Method, and is $\sim 2\times$ faster than BP.

Observation 3.22   Our FABP implementation is linear on the number of edges, with $\sim 2\times$ faster running time than BP on HADOOP.

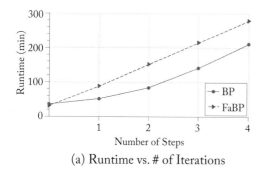 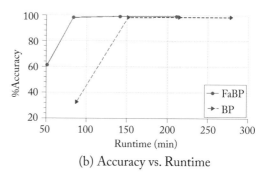

(a) Runtime vs. # of Iterations   (b) Accuracy vs. Runtime

Figure 3.8: Performance on the YahooWeb graph (best viewed in color): FABP wins on speed and wins/ties on accuracy. In (b), each of the method contains 4 points which correspond to the number of steps from 1–4. Notice that FABP achieves the maximum accuracy after 84 minutes, while BP achieves the same accuracy after 151 minutes.

PART II

Collective Graph Mining

CHAPTER 4

Summarization of Dynamic Graphs

Co-author: Neil Shah, Computer Science Department, Carnegie Mellon University— Chapter based on work that appeared at KDD 2015 [189].

In many applications, it is necessary or at least beneficial to explore multiple graphs collectively. These graphs can be temporal instances of the same set of objects (time-evolving graphs), or disparate networks coming from different sources. In this chapter we focus on dynamic networks: Given a large phonecall network over time, how can we describe it to a practitioner with just a few phrases? Other than the traditional assumptions about real-world graphs involving degree skewness, what can we say about their connectivity? For example, is the dynamic graph characterized by many large cliques which appear at fixed intervals of time, or perhaps by several large stars with dominant hubs that persist throughout? In this chapter we focus on these questions, and specifically on constructing concise summaries of large, real-world *dynamic* graphs in order to better understand their underlying behavior. In this chapter, we extend our work on single-graph summarization which we described in Chapter 2.

The problem of dynamic graph summarization has numerous practical applications. Dynamic graphs are ubiquitously used to model the relationships between various entities over *time*, which is a valuable feature in almost all applications in which nodes represent users or people [2, 21, 37, 82, 87, 135, 163, 203]. Examples include online social networks, phonecall networks, collaboration, coauthorship, and other interaction networks.

The traditional goals of clustering and community detection tasks (e.g., modularity-based community detection, spectral clustering, and cut-based partitioning) are not quite aligned with the endeavor we propose. These algorithms typically produce groupings of nodes which satisfy or approximate some optimization function. However, they do not offer a characterization of the outputs—are the detected groupings stars or chains, or perhaps dense blocks? Furthermore, the lack of explicit ordering in the groupings leaves a practitioner with limited time and no insights on where to begin understanding his data.

Here we propose TIMECRUNCH, an effective approach to concisely summarizing large, dynamic graphs which extend beyond traditional dense and isolated "caveman" communities. Similarly to the single-graph summarization work described in Chapter 2, we leverage MDL (Minimum Description Length) in order to find succinct graph patterns. In contrast to the static vocabulary we introduced before, in this chapter we seek to identify and appropriately describe graphs over time using a lexicon of *temporal phrases* which describe temporal connectivity behavior. Figure 4.1 shows results found from applying TimeCrunch to real-world dynamic graphs.

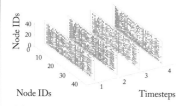 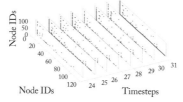 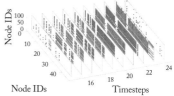

(a) 40 users of Yahoo! Messenger forming a *constant near clique* with unusually high 55% density, over 4 weeks in April 2008.

(b) 111 callers in a large phonecall network, forming a *periodic star*, over the last week of December 2007 (heavy activity on holidays).

(c) 43 collaborating biotech. Authors forming a *ranged near clique* in the DBLP network, jointly publishing through 2005-2012.

Figure 4.1: TimeCrunch finds coherent, interpretable temporal structures. We show the reordered subgraph adjacency matrices, over the timesteps of interest, each outlined in gray; edges are plotted in alternating red and blue, for discernability.

- Figure 4.1a shows a *constant near-clique* of 40 users in the Yahoo! messaging network interacting over 4 weeks in April 2008. The relevant subgraph has an unnaturally high 55% density over this duration. One possible explanation is that these users may be bots that message each other in an effort to appear normal and avoid suspension. We cannot verify, as the dataset is anonymized for privacy purposes.

- Figure 4.1b depicts a *periodic star* of 111 callers in the phonecall network of a large, anonymous Asian city during the last week of December 2007. We observe that the star behavior oscillates over time, and, specifically, odd-numbered timesteps have stronger star structure than the even-numbered ones. Furthermore, the appearance of the star is strongest on December 25th and 31st, corresponding to major holidays.

- Lastly, Figure 4.1c shows a *ranged near-clique* of 43 authors found in the DBLP network who jointly published in biotechnology journals, such as *Nature* and *Genome Research* from 2005–2012. This observation agrees with intuition as works in this field typically have many co-authors. The first and last timesteps shown have very sparse connectivity and were not part of the detected structure—they serve only to demarcate the range of activity.

In this chapter, we present a scalable solution to the following problem.

Problem 4.1 Dynamic Graph Summarization - Informal   Given a dynamic graph, i.e., a time sequence of adjacency matrices[1] $\mathbf{A}_1, \mathbf{A}_2, \ldots, \mathbf{A}_t$, find a set of possibly overlapping temporal subgraphs to **concisely describe** the given dynamic graph in a **scalable** fashion.

4.1   PROBLEM FORMULATION

In this section, we give the first main contribution of our work: formulation of dynamic graph summarization as a compression problem, using MDL. For clarity, in Table 4.1 we provide the recurrent symbols used in this chapter—for the reader's convenience, we repeat the definitions of symbols that we introduced in Chapter 2 for static graph summarization.

As a reminder to the reader, the Minimum Description Length (MDL) principle aims to be a practical version of Kolmogorov Complexity [138], often associated with the motto *Induction by Compression*. MDL states that given a model family $\mathcal{M}$, the best model $M \in \mathcal{M}$ for some observed data $\mathcal{D}$ is that which minimizes $L(M) + L(\mathcal{D}|M)$, where $L(M)$ is the length in bits used to describe M and $L(\mathcal{D}|M)$ is the length in bits used to describe $\mathcal{D}$ encoded using M. MDL enforces lossless compression for fairness in the model selection process.

For our application, we focus on the analysis of undirected dynamic graphs in tensor fashion using fixed-length, discretized time intervals. Our notation will reflect the treatment of the problem as one with a series of individual snapshots of graphs, or a tensor. Specifically, we consider a dynamic graph $G(\mathcal{V}, \mathcal{E})$ with $n = |\mathcal{V}|$ nodes, $m = |\mathcal{E}|$ edges and t timesteps, without self-loops. Here, $G = \cup_x G_x(\mathcal{V}, \mathcal{E}_x)$, where G_x and E_x correspond to the graph and edge-set for the xth timestep. The ideas proposed in this work, however, can easily be generalized to other types of dynamic graphs.

For our summary, we consider the set of temporal phrases $\Phi = \Delta \times \Omega$, where Δ corresponds to the set of temporal signatures, Ω corresponds to the set of static structure identifiers and $\times$ denotes the Cartesian set product. Although we can include arbitrary temporal signatures and static structure identifiers into these sets depending on the types of temporal subgraphs we expect to find in a given dynamic graph, we choose five temporal signatures which we anticipate to find in real-world dynamic graphs [18] : oneshot (o), ranged (r), periodic (p), flickering (f), and constant (c), and six very common structures found in real-world static graphs (Chapter 2) —stars (st), *full* and *near* cliques (fc, nc), *full* and *near* bipartite cores (bc, nb), and chains (ch). In summary, we have the signatures $\Delta = \{o, r, p, f, c\}$, static identifiers $\Omega = \{st, fc, nc, bc, nb, ch\}$ and temporal phrases $\Phi = \Delta \times \Omega$. We will further describe these signatures, identifiers and phrases after formalizing our objective.

---

[1]If the graphs have different, but overlapping node sets, $\mathcal{V}_1, \mathcal{V}_2, \ldots, \mathcal{V}_t$, we assume that $\mathcal{V} = \mathcal{V}_1 \cup \mathcal{V}_2 \cup \ldots \cup \mathcal{V}_t$, and the disconnected nodes are treated as singletons.

Table 4.1: TimeCrunch: Frequently used symbols and their definitions

Symbol	Description
$G, \mathbf{A}$	Dynamic graph and adjacency tensor, respectively
$G_x, \mathbf{A}_x$	x^{th} timestep, adjacency matrix of G, respectively
$\mathcal{E}_x, m_x$	Edge-set and number of edges of G_x, respectively
fc, nc	*Full* clique and *near* clique, respectively
fb, nb	*Full* bipartite core and *near* bipartite core, respectively
st	Star graph
ch	Chain graph
o	Oneshot
c	Constant
r	Ranged
p	Periodic
f	Flickering
t	Total number of timesteps for the dynamic graph
Δ	Set of temporal signatures
Ω	Set of static identifiers
Φ	Lexicon, set of temporal phrases $\Phi = \Delta \times \Omega$
$\times$	Cartesian set product
M, s	Model M, temporal structure $s \in M$, respectively
$\lvert S \rvert$	Cardinality of set S
$\lvert s \rvert$	# of nodes in structure s
$u(s)$	Timesteps in which structure s appears
$v(s)$	Temporal phrase of structure s, $v(s) \in \Phi$
$\mathbf{M}$	Approximation of $\mathbf{A}$ induced by M
$\mathbf{E}$	Error matrix $\mathbf{E} = \mathbf{M} \oplus \mathbf{E}$
$\oplus$	Exclusive OR
$L(G,M)$	# of bits used to encode M and G given M
$L(M)$	# of bits to encode M

In order to use MDL for dynamic graph summarization using these temporal phrases, we next define the model family $\mathcal{M}$, the means by which a model $M \in \mathcal{M}$ describes our dynamic graph and how to quantify the cost of encoding in terms of bits.

4.1.1   MDL FOR DYNAMIC GRAPH SUMMARIZATION

We consider models $M \in \mathcal{M}$ to be composed of ordered lists of temporal graph structures with node overlaps, but no edge overlaps. Each $s \in M$ describes a certain region of the adjacency tensor $\mathbf{A}$ in terms of the interconnectivity of its nodes. We will use $area(s, M, \mathbf{A})$ to describe the edges $(i, j, x) \in \mathbf{A}$ which s induces, writing only $area(s)$ when the context for M and $\mathbf{A}$ is clear.

Our model family $\mathcal{M}$ consists of all possible permutations of subsets of $\mathcal{C}$, where $\mathcal{C} = \cup_v \mathcal{C}_v$ and $\mathcal{C}_v$ denotes the set of all possible temporal structures of phrase $v \in \Phi$ over all possible combinations of timesteps. That is, $\mathcal{M}$ consists of all possible models M, which are ordered lists of temporal phrases $v \in \Phi$, such as flickering stars (*fst*), periodic full cliques (*pfc*), etc., over all possible subsets of $\mathcal{V}$ and $G_1 \cdots G_t$. Through MDL, we seek the model $M \in \mathcal{M}$ which minimizes the encoding length of the model M and the adjacency tensor $\mathbf{A}$ given M.

Our fundamental approach for transmitting the adjacency tensor $\mathcal{A}$ via the model M is described next. First, we transmit M. Next, given M, we induce the approximation of the adjacency tensor $\mathbf{M}$ as described by each temporal structure $s \in M$. For each structure s, we induce the edges in $area(s)$ in $\mathbf{M}$ accordingly. Given that $\mathbf{M}$ is a summary approximation to $\mathbf{A}$, $\mathbf{M} \neq \mathbf{A}$ most likely. Since MDL requires lossless encoding, we must also transmit the error $\mathbf{E} = \mathbf{M} \oplus \mathbf{A}$, obtained by taking the exclusive OR between $\mathbf{M}$ and $\mathbf{A}$. Given M and $\mathbf{E}$, a recipient can construct the full adjacency tensor $\mathbf{A}$ in a lossless fashion.

Thus, we formalize the problem we tackle as follows.

Problem 4.2   Minimum Dynamic Graph Description   Given a dynamic graph G with adjacency tensor $\mathbf{A}$ and temporal phrase lexicon Φ, find the smallest model M which minimizes the total encoding length

$$L(G, M) = L(M) + L(\mathbf{E}),$$

where $\mathbf{E} = \mathbf{M} \oplus \mathbf{A}$ is the error matrix and $\mathbf{M}$ is the approximation of $\mathbf{A}$ induced by M.

In the following subsections, we further formalize the task of encoding the model M and the error matrix $\mathbf{E}$.

4.1.2   ENCODING THE MODEL

The encoding length to fully describe a model $M \in \mathcal{M}$ is:

$$L(M) = \underbrace{L_{\mathbb{N}}(|M| + 1) + \log \binom{|M| + |\Phi| - 1}{|\Phi - 1|}}_{\text{\# of structures in total, and per type}} + \underbrace{\sum_{s \in M} (- \log P(v(s)|M) + L(c(s)) + L(u(s)))}_{\text{per structure: type, connectivity and temporal details}}.$$

(4.1)

We begin by transmitting the total number of temporal structures in M using $L_{\mathbb{N}}$, Rissanen's optimal encoding for integers greater than or equal to 1 [184]. Next, we optimally encode the number of temporal structures for each phrase $v \in \Phi$ in M. Then, for each structure s, we encode the type $v(s)$ for each structure $s \in M$ using optimal prefix codes [52], the connectivity $c(s)$ and the temporal presence of s, consisting of the ordered list of timesteps $u(s)$ in which s appears.

In order to have a coherent model encoding scheme, we next define the encoding for each phrase $v \in \Phi$ such that we can compute $L(c(s))$ and $L(u(s))$ for all structures in M. The connectivity $c(s)$ corresponds to the edges in $area(s)$ which are induced by s, whereas the temporal presence $u(s)$ corresponds to the timesteps in which s is present. We consider the connectivity and temporal presence separately, as the encoding for a temporal structure s described by a phrase v is the sum of encoding costs for the connectivity of the corresponding static structure identifier in Ω and its temporal presence as indicated by a temporal signature in Δ.

Encoding Connectivity

To compute the encoding cost $L(c(s))$ for the connectivity for each type of static structure identifier in our identifier set Ω (i.e., cliques, near-cliques, bipartite cores, near-bipartite cores, stars, and chains) we use the formulas introduced in Section 2.2.2.

Encoding Temporal Presence

For a given phrase $v \in \Phi$, it is not sufficient to only encode the connectivity of the underlying static structure. For each structure s, we must also encode the temporal presence $u(s)$, consisting of a set of ordered timesteps in which s appears. In this section, we describe how to compute the encoding cost $L(u(s))$ for each of the temporal signatures in the signature set Δ.

We note that describing a set of timesteps $u(s)$ in terms of temporal signatures in Δ is yet another model selection problem for which we can leverage MDL. As with connectivity encoding, labeling $u(s)$ with a given temporal signature may not be *precisely* accurate—however, any mistakes will add to the cost of transmitting the error. Errors in temporal presence encoding will be further detailed in Section 4.1.3.

Oneshot: Oneshot structures appear at only one timestep in $G_1 \cdots G_t$ – that is, $|u(s)| = 1$. These structures represent graph anomalies, in the sense that they are non-recurrent interactions which are only observed once. The encoding cost $L(o)$ for the temporal presence of a oneshot structure

o can be written as $L(o) = \log(t)$. As the structure occurs only once, we only have to identify the timestep of occurrence from the t observed timesteps.

Ranged: Ranged structures are characterized by a short-lived existence. These structures appear for several timesteps in a row before disappearing again—they are defined by a single burst of activity. The encoding cost $L(r)$ for a ranged structure r is given by:

$$L(r) \quad = \quad \underbrace{L_{\mathbb{N}}(|u(s)|)}_{\text{\# of timesteps}} + \quad \underbrace{\log\binom{t}{2}}_{\text{start and end timestep IDs}} . \tag{4.2}$$

We first encode the number of timesteps in which the structure occurs, followed by the timestep IDs of both the start and end timestep marking the span of activity.

Periodic: Periodic structures are an extension of ranged structures in that they appear at fixed intervals. However, these intervals are spaced greater than one timestep apart. As such, the same encoding cost function we use for ranged structures suffices here. That is, $L(p)$ for a periodic structure p is given by $L(p) = L(r)$.

For both ranged and periodic structures, periodicity can be inferred from the start and end markers along with the number of timesteps $|u(s)|$, allowing the reconstruction of the original $u(s)$.

Flickering: A structure is flickering if it appears only in some of the $G_1 \cdots G_t$ timesteps, and does so without any discernible ranged/periodic pattern. The encoding cost $L(f)$ for a flickering structure f is as follows:

$$L(f) \quad = \quad \underbrace{L_{\mathbb{N}}(|u(s)|)}_{\text{\# of timesteps}} + \quad \underbrace{\log\binom{n}{|u(s)|}}_{\text{IDs for the timesteps of occurrence}} .$$

We encode the number of timesteps in which the structure occurs in addition to the IDs for the timesteps of occurrence.

Constant: Constant structures persist throughout all timesteps. That is, they occur at each timestep $G_1 \cdots G_t$ without exception. In this case, our encoding cost $L(c)$ for a constant structure c is defined as $L(c) = 0$. Intuitively, information regarding the timesteps in which the structure appears is "free," as it is already given by encoding the phrase descriptor $v(s)$.

4.1.3   ENCODING THE ERRORS

Given that M is a summary and the $\mathbf{M}$ induced by M is only an approximation of $\mathbf{A}$, it is necessary to encode errors made by M. In particular, there are two types of errors we must consider. The first is error in connectivity—that is, if $area(s)$ induced by structure s is not *exactly* the same as the associated patch in $\mathbf{A}$, we encode the relevant mistakes. The second is the error induced by encoding the set of timesteps $u(s)$ with a fixed temporal signature, given that $u(s)$ may not precisely follow the temporal pattern used to encode it.

Encoding Errors in Connectivity

Following the same principles described in Section 2.2.3, we encode the error tensor $\mathbf{E} = \mathbf{M} \oplus \mathbf{A}$ as two different pieces: $\mathbf{E}^+$ and $\mathbf{E}^-$. The first piece, $\mathbf{E}^+$, refers to the area of $\mathbf{A}$ which M models and the area of $\mathbf{M}$ that includes extraneous edges not present in the original graph. The second piece, $\mathbf{E}^-$, consists of the area of $\mathbf{A}$ which M does not model and therefore does not describe. As a reminder, the encoding for $\mathbf{E}^+$ and $\mathbf{E}^-$ is:

$$
\begin{aligned}
L(\mathbf{E}^+) &= \log(|\mathbf{E}^+|) + ||\mathbf{E}^+||l_1 + ||\mathbf{E}^+||'l_0 \\
L(\mathbf{E}^-) &= \log(|\mathbf{E}^-|) + ||\mathbf{E}^-||l_1 + ||\mathbf{E}^-||'l_0,
\end{aligned}
$$

$$\underbrace{\phantom{\log(|\mathbf{E}^-|)}}_{\text{\# of edges}} \quad \underbrace{\phantom{||\mathbf{E}^-||l_1 + ||\mathbf{E}^-||'l_0}}_{\text{edges}}$$

where $||E||$ and $||E||'$ denote the counts for existing and non-existing edges in $area(E)$, respectively. Then, $l_1 = -\log(||E||/(||E|| + ||E||'))$ and $l_0 = -\log(||E||'/(||E|| + ||E||'))$ represent the length of the optimal prefix codes for the existing and non-existing edges, respectively. For more explanations about our choices, refer to Section 2.2.3.

Encoding Errors in Temporal Presence

For encoding errors induced by identifying $u(s)$ as one of the temporal signatures, we turn to optimal prefix codes applied over the error distribution for each structure s. Given the information encoded for each signature type in Δ, we can reconstruct an approximation $\tilde{u}(s)$ of the original timesteps $u(s)$ such that $|u(s)| = |\tilde{u}(s)|$. Using this approximation, the encoding cost $L(e_u(s))$ for the error $e_u(s) = u(s) - \tilde{u}(s)$ is defined as

$$
L(e_u(s)) = \sum_{k \in h(e_u(s))} \left(\underbrace{\log(k)}_{\text{error magnitude}} + \underbrace{\log c(k)}_{\text{\# of occurrences}} + \underbrace{c(k)l_k}_{\text{error}} \right),
$$

where $h(e_u(s))$ denotes the set of elements with unique magnitude in $e_u(s)$, $c(k)$ denotes the count of element k in $e_u(s)$, and l_k denotes the length of the optimal prefix code for k. For each magnitude error, we encode the magnitude of the error, the number of times it occurs, and the actual errors using optimal prefix codes. Using the model in conjunction with the temporal presence and connectivity errors, a recipient can first recover the $u(s)$ for each $s \in M$, approximate $\mathbf{A}$ with $\mathbf{M}$ induced by M, produce $\mathbf{E}$ from $\mathbf{E}^+$ and $\mathbf{E}^-$, and finally recover $\mathbf{A}$ losslessly through $\mathbf{A} = \mathbf{M} \oplus \mathbf{E}$.

Remark: For a dynamic graph G of n nodes, the search space $\mathcal{M}$ for the best model $M \in \mathcal{M}$ is intractable, as it consists of all the permutations of all possible temporal structures over the lexicon Φ, over all possible subsets over the node-set $\mathcal{V}$ and over all possible graph timesteps $G_1 \cdots G_t$. Furthermore, $\mathcal{M}$ is not easily exploitable for efficient search. Thus, we propose several practical approaches for the purpose of finding good and interpretable temporal models/summaries for G.

4.2   TIMECRUNCH: VOCABULARY-BASED SUMMARIZATION OF DYNAMIC GRAPHS

Thus far, we have described our strategy of formulating dynamic graph summarization as a problem in a compression context for which we can leverage MDL. Specifically, we have detailed how to encode a model and the associated error which can be used to losslessly reconstruct the original dynamic graph G. Our models are characterized by ordered lists of temporal structures which are further classified as *phrases* from the lexicon Φ. Each $s \in M$ is identified by a phrase $p \in \Phi$ over

- the node connectivity $c(s)$, i.e., an induced set of edges depending on the static structure identifier st, fc, etc.), and
- the associated temporal presence $u(s)$, i.e., an ordered list of timesteps captured by a temporal signature o, r, etc. and deviations) in which the temporal structure is active.

The error consists of the edges which are not covered by $\mathbf{M}$, the approximation of $\mathbf{A}$ induced by M.

Next, we discuss how we find good candidate temporal structures to populate the candidate set $\mathcal{C}$, as well as how we find the best model M with which to summarize our dynamic graph. The pseudocode for our algorithm is given in Algorithm 4.1 and the next subsections detail each step of our approach.

4.2.1   GENERATING CANDIDATE STATIC STRUCTURES

TIMECRUNCH takes an incremental approach to dynamic graph summarization. That is, our approach begins with considering potentially useful subgraphs for summarization over the static graphs $G_1 \cdots G_t$. There are numerous static graph decomposition algorithms which enable community detection, clustering and partitioning for these purposes, such as EigenSpokes [176], METIS [108], spectral partitioning [13], Graclus [54], cross-associations [40], Subdue [113], and SlashBurn [103]. Summarily, for each $G_1 \ldots G_t$, a set of subgraphs $\mathcal{F}$ is produced.

Algorithm 4.1 TIMECRUNCH

Input : input dynamic graphs from time 1 to time t: $G_1 \cdots G_t$
Output : summarization M of dynamic graph G

1: **Generating Candidate Static Subgraphs**: Generate static subgraphs for each $G_1 \cdots G_t$ using traditional static graph decomposition approaches.
2: **Labeling Candidate Static Subgraphs**: Label each static subgraph as a static structure corresponding to the identifier $x \in \Omega$ which minimizes the *local encoding cost*.
3: **Stitching Candidate Temporal Structures**: *Stitch* the static structures from $G_1 \ldots G_t$ together to form temporal structures with coherent connectivity behavior, and label them according to the phrase $p \in \Phi$ which minimizes the temporal presence encoding cost.
4: **Composing the Summary**: Compose a model M of important, non-redundant temporal structures which summarize G using the VANILLA, TOP10, TOP-100 and STEPWISE heuristics. Choose M associated with the heuristic that produces the smallest total encoding cost.

4.2.2   LABELING CANDIDATE STATIC STRUCTURES

Once we have the set of static subgraphs from $G_1 \ldots G_t$, $\mathcal{F}$, we next seek to label each subgraph in $\mathcal{F}$ according to the static structure identifiers in Ω that best fit the connectivity for the given subgraph.

Definition 4.3   Static structures.   A static structure is a static subgraph that is labeled with a static identifier in $\Omega = \{fc, nc, fb, nb, st, ch\}$.

For each subgraph construed as a set of nodes $\mathcal{L} \in \mathcal{V}$ for a fixed timestep, does the adjacency matrix of $\mathcal{L}$ best resemble a star, near or full clique, near or full bipartite core or a chain? To answer this question, we leverage the encoding scheme discussed in Section 2.3.2. In a nutshell, we try encoding the subgraph $\mathcal{L}$ using each of the static identifiers in Ω and label it with the identifier $x \in \Omega$ which minimizes the encoding cost.

4.2.3   STITCHING CANDIDATE TEMPORAL STRUCTURES

Thus far, we have a set of static subgraphs $\mathcal{F}$ over $G_1 \ldots G_t$ labeled with the associated static identifiers which best represent the subgraph connectivity (from now on, we refer to $\mathcal{F}$ as a set of static *structures* instead of *subgraphs* as they have been labeled with identifiers). From this set, our goal is to find temporal structures—namely, we seek to *find* static subgraphs which have the same patterns of connectivity over one or more timesteps and *stitch* them together. Thus, we formulate the problem of finding coherent temporal structures in G as a clustering problem over

$\mathcal{F}$. Although there are several criteria that we could use for clustering static structures together, we employ the following based on their intuitive meaning.

Definition 4.4 Temporal structures.   Two static structures belong to the same temporal structure (i.e., they are in the same cluster) if they have

- substantial overlap in the node-sets composing their respective subgraphs, and
- exactly the same, or similar (full and near clique, or full and near bipartite core) static structure identifiers.

These criteria, if satisfied, allow us to find groups of nodes that share interesting connectivity patterns over time. For example, in a phonecall network, the nodes "Smith," "Johnson," and "Tompson" who call each other every Sunday form a periodic clique (temporal structure).

Conducting the clustering by naively comparing each static structure in $\mathcal{F}$ to the others will produce the desired result, but is *quadratic* on the number of static structures and is thus undesirable from a scalability point of view. Instead, we propose an incremental approach using repeated rank-1 Singular Value Decomposition (SVD) for clustering the static structures, which offers *linear* time complexity on the number of edges m in G.

Matrix definitions for clustering   We first define the matrices that we will use to cluster the static structures.

Definition 4.5 SNMM.   The structure-node membership matrix (SNMM), $\mathbf{B}$, is a $|\mathcal{F}| \times |\mathcal{V}|$ matrix, where $\mathbf{B}_{ij}$ indicates whether the ith row (structure) in $\mathcal{F}$ ($\mathbf{B}$) contains node j in its node-set. Thus, $\mathbf{B}$ is a matrix indicating the membership of the nodes in $\mathcal{V}$ to each of the static structures in $\mathcal{F}$.

We note that any two equivalent rows in $\mathbf{B}$ are characterized by structures that share the same node-set (but possibly different static identifiers). As our clustering criteria mandate that we only cluster structures with the same or similar static identifiers, in our algorithm, we construct four SNMMs—$\mathbf{B}_{st}$, $\mathbf{B}_{cl}$, $\mathbf{B}_{bc}$, and $\mathbf{B}_{ch}$ corresponding to the associated matrices for stars, near and full cliques, near and full bipartite cores and chains, respectively. Now, any two equivalent rows in $\mathbf{B}_{cl}$ are characterized by structures that share the same node-set, and the same or similar static identifiers (full or near-clique), and analogue for the other matrices. Next, we utilize SVD to cluster the rows in each SNMM, effectively clustering the structures in $\mathcal{F}$.

Clustering with SVD   We first give the definition of SVD, and then describe how we can use it to cluster the static structures and discover temporal structures.

Definition 4.6 SVD.   The rank-k SVD of an $m \times n$ matrix $\mathbf{A}$ factorizes it into three matrices: the $m \times k$ matrix of left-singular vectors $\mathbf{U}$, the $k \times k$ diagonal matrix of singular values $\mathbf{\Sigma}$, and the $n \times k$ matrix of right-singular vectors $\mathbf{V}$, such that $\mathbf{A} = \mathbf{U}\mathbf{\Sigma}\mathbf{V}^{\mathsf{T}}$.

A rank-k SVD effectively reduces the input data to the best k-dimensional representation, each of which can be mined separately for clustering and community detection purposes. However, one major issue with using SVD in this fashion is that identifying the desired number of clusters k upfront is a non-trivial task. To this end, [169] evidences that in cases where the input matrix is sparse, repeatedly clustering using k rank-1 decompositions and adjusting the input matrix accordingly approximates the batch rank-k decomposition. This is a valuable result in our case. As we do not initially know the number of clusters needed to group the structures in $\mathcal{F}$, we eliminate the need to define k altogether by repeatedly applying rank-1 SVD using power iteration and removing the discovered clusters from each SNMM until all clusters have been found (when all SNMMs are fully sparse and thus *deflated*). However, in practice, full deflation is unnecessary for summarization purposes, as the most dominant clusters are found in early iterations due to the nature of SVD. For each of the SNMMs, the matrix $\mathbf{B}$ used in the $(i+1)^{th}$ iteration of this iterative process is computed as

$$\mathbf{B}^{i+1} = \mathbf{B}^i - I^{\mathcal{G}_i} \circ \mathbf{B}^i,$$

where $\mathcal{G}_i$ denotes the set of row IDs corresponding to the structures which were clustered together in iteration i, $I^{\mathcal{G}_i}$ denotes the indicator matrix with 1s in rows specified by $\mathcal{G}_i$ and $\circ$ denotes the Hadamard matrix product. This update to $\mathbf{B}$ is needed between iterations, as without subtracting the previously found cluster, repeated rank-1 decompositions would find the same cluster ad infinitum, and the algorithm would not converge.

Although this algorithm works assuming that we can remove a cluster in each iteration, the question of how we find this cluster given a singular vector has yet to be answered. First, we sort the singular vector, permuting the rows by magnitude of projection. The intuition is that the structure (rows) which projects most strongly to that cluster is the best representation of the cluster, and is considered a *base* structure which we attempt to find matches for. Starting from the base structure, we iterate down the sorted list and compute the Jaccard similarity, defined as $J(\mathcal{L}_1, \mathcal{L}_2) = \frac{|\mathcal{L}_1 \cap \mathcal{L}_2|}{|\mathcal{L}_1 \cup \mathcal{L}_2|}$ for node-sets $\mathcal{L}_1$ and $\mathcal{L}_2$, between each structure and the base. Other structures which are composed of the same, or similar node-sets will also project strongly to the cluster, and will be stitched to the base. Once we encounter a series of structures which fail to match by a predefined similarity criterion, we adjust the SNMM and continue with the next iteration.

Having stitched together the relevant static structures, we label each temporal structure using the temporal signature in Δ and resulting phrase in Φ which minimizes its encoding cost using the temporal encoding framework derived in Section 4.1.2. We use these temporal structures to populate the candidate set $\mathcal{C}$ for our model.

4.2.4   COMPOSING THE SUMMARY

Given the candidate set of temporal structures $\mathcal{C}$, we next seek to find the model M which best summarizes G. However, actually finding the best model is combinatorial, as it involves considering all possible permutations of subsets of C and choosing the one which gives the smallest encoding cost. As a result, we propose several heuristics that give fast and approximate solutions without entertaining the entire search space. As in the case of static graphs in Chapter 2, to reduce the search space, we associate a metric with each temporal structure by which we measure quality, called the *local encoding benefit*. The local encoding benefit is defined as the ratio between the cost of encoding the given temporal structure as error, and the cost of encoding it using the best phrase (local encoding cost). Large local encoding benefits indicate high compressibility, and thus an easy-to-describe structure in the underlying data. We use the same heuristics we used in static graph summarization (Chapter 2).

Plain: This is the baseline approach, in which our summary contains all the structures from the candidate set, or $M = C$.

Top-k: In this approach, M consists of the top k structures of C, sorted by local encoding benefit.

Greedy'nForget: This approach involves considering each structure of C, sorted by local encoding benefit, and adding it to M if the global encoding cost decreases. If adding the structure to M increases the global encoding cost, the structure is discarded as redundant or not worthwhile for summarization purposes.

In practice, TimeCrunch uses each of the heuristics and identifies the best summary for G as the one that produces the minimum encoding cost.

4.3   EMPIRICAL RESULTS

In this section, we evaluate TimeCrunch and seek to answer the following questions: Are real-world dynamic graphs well structured, or noisy and indescribable? If they are structured, what temporal structures do we see in these graphs and what do they mean? Lastly, is TimeCrunch scalable?

Datasets and Experimental Setup.   For our experiments, we use five real dynamic graph datasets, which are summarized in Table 4.2 and described below.

Table 4.2: Dynamic graphs used for empirical analysis

Graph	Nodes	Edges	Timesteps
Enron [202]	151	20,000	163 weeks
Yahoo-IM [233]	100,000	2.1 million	4 weeks
Honeynet	372,000	7.1 million	32 days
DBLP [55]	1.3 million	15 million	25 years
Phonecall	6.3 million	36.3 million	31 days

Enron: The `Enron` e-mail dataset is publicly available. It contains 20,000 unique links between 151 users based on e-mail correspondence over 163 weeks (May 1999–June 2002).

• **Yahoo! IM**: The `Yahoo-IM` dataset is publicly available. It contains 2.1 million sender-receiver pairs between 100,000 users over 5,709 zip codes selected from the Yahoo! messenger network over 4 weeks starting from April 1, 2008.

• **Honeynet**: The `Honeynet` dataset contains information about network attacks on *honeypots* (i.e., computers which are left intentionally vulnerable to attackers). It contains source IP, destination IP and attack timestamps of 372 thousand (attacker and honeypot) machines with 7.1 million unique daily attacks over a span of 32 days starting from December 31, 2013.

• **DBLP**: The `DBLP` computer science bibliography is publicly available, and contains yearly co-authorship information, indicating joint publication. We used a subset of DBLP spanning 25 years, from 1990–2014, with 1.3 million authors and 15 million unique author-author collaborations over the years.

• **Phonecall**: The `Phonecall` dataset describes the who-calls-whom activity of 6.3 million individuals from a large, anonymous Asian city and contains a total of 36.3 million unique daily phonecalls. It spans 31 days, starting from December 1, 2007.

In our experiments, we use SLASHBURN [103] for generating candidate static structures, as it is scalable and designed to extract structure from real-world, non-caveman graphs as it is scalable and designed to extract structure from real-world, non-caveman graphs.[2] We note that, thanks to our MDL approach, the inclusion of additional graph decomposition methods will only *improve* the results. Furthermore, when clustering each sorted singular vector during the stitching process, we move on with the next iteration of matrix deflation after 10 failed matches with a Jaccard similarity threshold of 0.5—we choose 0.5 based on experimental results which show that it gives the best encoding cost and balances between excessively terse and overlong (error-prone) models. Lastly, we run TIMECRUNCH for a total of 5,000 iterations for all graphs (each iteration uniformly selects one SNMM to mine, resulting in 5,000 total temporal structures), except for the `Enron` graph which is fully deflated after 563 iterations and the `Phonecall` graph which we limit to 1,000 iterations for efficiency.

4.3.1   QUANTITATIVE ANALYSIS

In this section, we use TIMECRUNCH to summarize each of the real-world dynamic graphs from Table 4.2 and report the resulting encoding costs. Specifically, the evaluation is done by comparing the compression ratio between the encoding costs of the resulting models to the null encoding (ORIGINAL) cost, which is obtained by encoding the graph using an empty model.

We note that although we provide results in a compression context, as in the case of static graph summarization, compression is *not* our main goal for TIMECRUNCH, but rather the means

---

[2]A caveman graph arises by modifying a set of fully connected clusters (caves) by removing one edge from each cluster and using it to connect to a neighboring one such that the clusters form a single loop [220].

Table 4.3: TimeCrunch finds temporal structures that can compress real graphs. Original denotes the cost in bits for encoding each graph with an empty model. Columns under Time-Crunch show relative costs for encoding the graphs using the respective heuristic (size of model is parenthesized). The lowest description cost is bolded.

Graph	Original (bits)	TimeCrunch			
		Vanilla	Top 10	Top 100	Greedy'nForget
Enron	86,102	89% (563)	88%	81%	78% (130)
Yahoo-IM	16,173,388	97% (5,000)	99%	98%	93% (1,523
Honeynet	72,081,235	82% (5,000)	96%	89%	81% (3,740
DBLP	167,831,004	97% (5,000)	99%	99%	96% (1,627)
Phonecall	478,377,701	100% (1,000)	100%	99%	98% (370)

to our end for identifying suitable structures with which to summarize dynamic graphs and route the attention of practitioners. For this reason, we do not evaluate against other, compression-oriented methods which prioritize leveraging any correlation within the data to reduce cost and save bits. Other temporal clustering and community detection approaches which focus only on extracting dense blocks are also not compared to our method for similar reasons.

In our evaluation, we consider (a) Original and (b) TimeCrunch summarization using the proposed heuristics. In the Original approach, the entire adjacency tensor is encoded using the empty model $M = \emptyset$. As the empty model does not describe any part of the graph, all the edges are encoded using $L(\mathbf{E}^-)$. We use this as a baseline to evaluate the savings attainable using TimeCrunch. For summarization using TimeCrunch, we apply the Vanilla, Top10, Top-100, and Greedy'nForget model selection heuristics. We note that we ignore very small structures of less than five nodes for Enron and less than eight nodes for the other, larger datasets.

Table 4.3 shows the results of our experiments in terms of the encoding costs of various summarization techniques as compared to the Original approach. Smaller compression ratios indicate better summaries, with more structure explained by the respective models. For example, Greedy'nForget was able to encode the Enron dataset using just 78% of the bits compared to 89% using Vanilla. In our experiments, we find that the Greedy'nForget heuristic produces models with considerably fewer structures than Vanilla, while giving even more concise graph summaries (Figure 4.2). This is because it is highly effective in pruning redundant, overlapping or error-prone structures from the candidate set $\mathcal{C}$, by evaluating new structures in the context of previously seen ones.

Observation 4.7   Real-world dynamic graphs are structured. TimeCrunch gives a better encoding cost than Original, indicating the presence of a temporal graph structure.

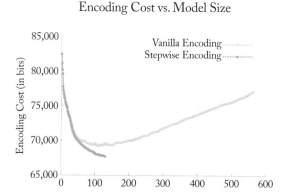

Figure 4.2: TimeCrunch-Greedy'nForget summarizes Enron using just 78% of Original's bits and 130 structures compared to 89% and 563 structures of TimeCrunch-Vanilla by pruning unhelpful structures from the candidate set.

4.3.2   QUALITATIVE ANALYSIS

In this section, we discuss qualitative results from applying TimeCrunch to the graphs mentioned in Table 4.2.

Enron   The Enron graph is mainly characterized by many periodic, ranged, and oneshot stars and several periodic and flickering cliques. Periodicity is reflective of office e-mail communications (e.g., meetings, reminders). Figure 4.3a shows an excerpt from one flickering clique which corresponds to several members of Enron's legal team, including Tana Jones, Susan Bailey, Marie Heard, and Carol Clair—all lawyers at Enron. Figure 4.3b shows an excerpt from a flickering star, corresponding to many of the same members as the flickering clique—the center of this star was identified as the boss, Tana Jones (Enron's Senior Legal Specialist)—note the vertical points above node 1 correspond to the satellites of the star and oscillate over time. Interestingly, the flickering star and clique extend over most of the observed duration. Furthermore, several of the oneshot stars corresponds to company-wide emails sent out by key players John Lavorato (CEO of Enron America), Sally Beck (COO), and Kenneth Lay (CEO/Chairman).

Yahoo! IM   The Yahoo-IM graph is composed of many temporal stars and cliques of all types, and several smaller bipartite cores with just a few members on one side (indicative of friends who share mostly similar friend-groups but are themselves unconnected). We observe several interesting patterns in this data. Figure 4.3d corresponds to a constant star with a hub that communicates with 70 users consistently over 4 weeks. We suspect that these users are part of a small office network, where the boss uses group messaging to notify employees of important updates or events—we notice that very few edges of the star are missing each week and the average degree of the satellites is roughly 4, corresponding to possible communication between employees. Figure 4.3c depicts a constant clique between 40 users, with an average density over 55%—we suspect that these may be spam-bots messaging each other in an effort to appear normal, or a large group of friends with multiple message groups. Due to lack of ground-truth, we cannot verify.

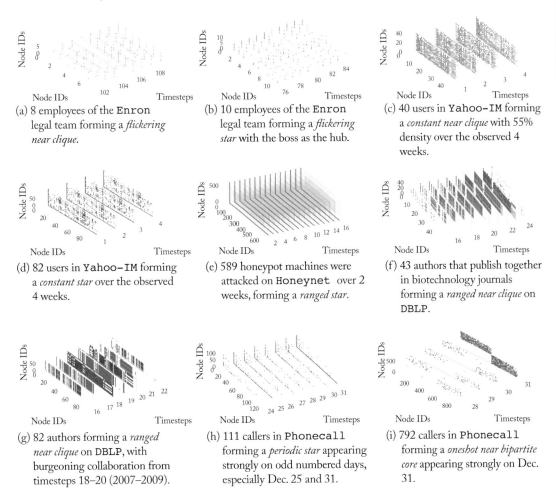

(a) 8 employees of the `Enron` legal team forming a *flickering near clique*.

(b) 10 employees of the `Enron` legal team forming a *flickering star* with the boss as the hub.

(c) 40 users in `Yahoo-IM` forming a *constant near clique* with 55% density over the observed 4 weeks.

(d) 82 users in `Yahoo-IM` forming a *constant star* over the observed 4 weeks.

(e) 589 honeypot machines were attacked on `Honeynet` over 2 weeks, forming a *ranged star*.

(f) 43 authors that publish together in biotechnology journals forming a *ranged near clique* on `DBLP`.

(g) 82 authors forming a *ranged near clique* on `DBLP`, with burgeoning collaboration from timesteps 18–20 (2007–2009).

(h) 111 callers in `Phonecall` forming a *periodic star* appearing strongly on odd numbered days, especially Dec. 25 and 31.

(i) 792 callers in `Phonecall` forming a *oneshot near bipartite core* appearing strongly on Dec. 31.

Figure 4.3: TIMECRUNCH finds meaningful temporal structures in real graphs. We show the reordered subgraph adjacency matrices over multiple timesteps. Individual timesteps are outlined in gray, and edges are plotted with alternating red and blue color for discernibility.

Honeynet   As we mentioned above, `Honeynet` is a bipartite graph between attacker and honeypot (victim) machines. As such, it is characterized by temporal stars and bipartite cores. Many of the attacks only span a single day, as indicated by the presence of 3,512 oneshot stars, and no attacks span the entire 32 day duration. Interestingly, 2,502 of these oneshot star attacks (71%) occur on the first and second observed days (December 31st and January 1st) indicating intentional "new-year" attacks. Figure 4.3e shows a ranged star, lasting 15 consecutive days and targeting 589 machines for the entire duration of the attack (node 1 is the hub of the star and the remainder are satellites).

DBLP   Agreeing with intuition, DBLP consists of a large number of oneshot temporal structures corresponding to many single instances of joint publication. However, we also find numerous ranged/periodic stars and cliques which indicate coauthors publishing in consecutive years or intermittently. Figure 4.3f shows a ranged clique spanning from 2007–2012 between 43 coauthors who jointly published each year. The authors are mostly members of the NIH NCBI (National Institute of Health National Center for Biotechnology Information) and have published their work in various biotechnology journals, such as *Nature, Nucleic Acids Research*, and *Genome Research*. Figure 4.3g shows another ranged clique from 2005–2011, consisting of 83 coauthors who jointly publish each year, with an especially collaborative 3 years (timesteps 18–20) corresponding to 2007–2009 before returning to status quo.

Phonecall   The Phonecall dataset is largely comprised of temporal stars and few dense clique and bipartite structures. Again, we have a large proportion of oneshot stars which occur only at single timesteps. Further analyzing these results, we find that 111 of the 187 oneshot stars (59%) are found on December 24th, 25th, and 31st, corresponding to Christmas Eve/Day and New Year's Eve holiday greetings. Furthermore, we find many periodic and flickering stars typically consisting of 50–150 nodes, which may be associated with businesses regularly contacting their clientele, or public phones which are used consistently by the same individuals. Figure 4.3h shows one such periodic star of 111 users over the last week of December, with particularly clear star structure on December 25th and 31st and other odd-numbered days, accompanied by substantially weaker star structure on the even-numbered days. Figure 4.3i shows an oddly well-separated oneshot near-bipartite core which appears on December 31st, consisting of two roughly equal-sized parts of 402 and 390 callers. Though we do not have ground truth to interpret these structures, we note that a practitioner with the appropriate information could better interpret their meaning.

4.3.3   SCALABILITY

All components of TIMECRUNCH (candidate subgraph generation, static subgraph labeling, temporal stitching and summary composition) are carefully designed to be near-linear on the number of edges. Figure 4.4 shows the $O(m)$ runtime of TIMECRUNCH on several induced temporal subgraphs (up to 14 M edges) taken from the DBLP dataset at varying time-intervals. We ran the experiments using a machine with 80 Intel Xeon(R) 4850 2 GHz cores and 256 GB

RAM. We use MATLAB for candidate subgraph generation and temporal stitching, and Python for model selection heuristics.

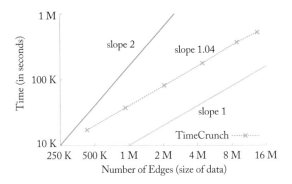

Figure 4.4: Runtime vs. data size. TimeCrunch scales near-linearly on the number of edges in the graph. Here, we use several induced temporal subgraphs from DBLP, up to 14 M edges in size.

4.4   RELATED WORK

The related work falls into three main categories: static graph mining, temporal graph mining, and graph compression and summarization. Table 4.4 gives a visual comparison of Time-Crunch with existing methods.

Static Graph Mining   Most works find specific, tightly knit structures, such as (near-) cliques and bipartite cores: eigendecomposition [188], cross-associations [40], and modularity-based optimization methods [28, 160]. Dhillon et al. [55] propose information-theoretic co-clustering based on mutual information optimization. However, these approaches have limited vocabularies and are unable to find other types of interesting structures, such as stars or chains. [54, 109] propose cut-based partitioning, whereas [13] suggests spectral partitioning using multiple eigenvectors—these schemes seek hard clustering of all nodes as opposed to identifying communities, and are not usually parameter-free. Subdue [50] and other fast frequent-subgraph mining algorithms [100] operate on labeled graphs. TimeCrunch involves unlabeled graphs and lossless compression. Analysis of multi-layer networks (networks in which the entities participate in multiple relationships) has been studied in the physics community [30, 115], although the focus is not on summarization.

Temporal Graph Mining   Most work on temporal graphs focuses on the evolution of specific properties [21, 87], change detection (e.g., using projected clustering [3]), or community detection. For example, GraphScore [201] and Com2 [18] use graph-search and PARAFAC (or Canonical Polyadic—CP) tensor decomposition followed by MDL to find dense tem-

Table 4.4: Feature-based comparison of TimeCrunch with alternative approaches

	Temporal	Time-consec.	Time-agnostic	Dense Blocks	Stars	Chains	Interpretable	Scalable	Prm-free
GraphScope [210]	✓	✓	✗	✓	✗	✗	✗	✓	✓
Com2 [18]	✓	✓	✓	✓	✓	✗	✗	✓	✗
Graph partitioning [13, 57, 115]	✗	✗	✗	✓	✗	✗	✗	✓	✗
Community detection [29, 168, 196]	✗	✗	✗	✓	✗	✗	✗	?	?
VoG [131], Chap. 2	✗	✗	✗	✓	✓	✓	✓	✓	✓
TimeCrunch	✓	✓	✓	✓	✓	✓	✓	✓	✓

poral cliques and bipartite cores. [68] uses incremental cross-association for change detection in dense blocks over time, whereas [174] proposes an algorithm for mining cross-graph quasi-cliques (though not in a temporal context). A probabilistic approach based on mixed-membership blockmodels is proposed by Fu et al. [74]. Significant efforts have focused on characterizing temporal networks with the frequency of their network motifs or recurring patterns [97, 129, 170, 234]. These approaches have limited vocabularies and/or do not offer temporal interpretability. Dynamic clustering [223] aims to find stable clusters over time by penalizing deviations from incremental static clustering. TimeCrunch focuses on interpretable structures, which may not appear at every timestep.

Graph Compression and Summarization   Work on summarization and compression of time-evolving graphs is quite limited [141]. Examples for compressing *static* graphs include Slash-Burn [103], which is a recursive node-reordering approach to leverage run-length encoding for graph compression; weighted graph compression [208] that uses structural equivalence to collapse nodes/edges to simplify graph representation; and others that we discussed extensively in Section 2.6. These approaches though do not readily operate on dynamic graphs. Moreover, VoG (Chapter 2, [123, 124]) uses MDL to label subgraphs in terms of a vocabulary on static graphs, consisting of stars, (near) cliques, (near) bipartite cores and chains. This approach only applies to static graphs and does not offer a clear extension to dynamic graphs. This chapter introduces a suitable lexicon for dynamic graphs, uses MDL to label *temporally coherent* subgraphs and proposes an effective and scalable algorithm for finding them. More recent works on time-evolving networks include graph stream summarization [204] for query efficiency (based on sketching), and influence-based graph summarization [1, 178, 194], which aim at summarizing influence or propagation processes in social and other networks.

CHAPTER 5

Graph Similarity

A question that often comes up when studying multiple networks is: How much do two graphs or networks differ in terms of connectivity, and which are the main node and edge culprits for their difference? For example, how much has a network changed since yesterday? How different is the wiring of Bob's brain (a left-handed male) and Alice's brain (a right-handed female), and what are their main differences?

Similarity or comparison of aligned graphs (i.e., with known node correspondence) is a core task for sense-making: abnormal changes in network traffic may indicate a computer attack; differences of big extent in a who-calls-whom graph may reveal a national celebration, or a telecommunication problem. Besides, network similarity can serve as a building block for similarity-based classification [45] of graphs, and give insights into transfer learning and behavioral patterns: is the Facebook message graph similar to the Facebook wall-to-wall graph? Tracking changes in networks over time, spotting anomalies and detecting events is a research direction that has attracted much interest (e.g., [39, 161, 219]).

Long in the purview of researchers, graph similarity has been a well-studied problem and several approaches have been proposed to solve variations of the problem. However, graph comparison with node/edge attribution still remains an open problem, while (with the passage of time) its list of requirements increases: the exponential growth of graphs, both in number and size, calls for methods that are not only accurate, but also scalable to graphs with billions of nodes.

In this chapter, we address three main problems: (i) How to compare two networks efficiently, (ii) how to evaluate the degree of their similarity, and (iii) how to identify the culprit nodes/edges responsible for the differences. We present DELTACON (for "δ connectivity" change detection), a principled, intuitive and scalable graph similarity method, and use it in real-world applications, such as temporal anomaly detection and graph clustering/classification. Table 5.1 gives the major symbols we use in the chapter and their definitions.

5.1 INTUITION

How can we find the similarity in connectivity between two graphs or, more formally, how can we solve the following problem?

Table 5.1: DELTACON: Symbols and definitions. Bold uppercase letters: matrices; bold lowercase letters: vectors; plain font: scalars.

Symbol	Description
$sim(G_1, G_2)$	Similarity between graphs G_1 and G_2
$d(G_1; G_2)$	Distance between graphs G_1 and G_2
S	$n \times n$ matrix of final scores with elements s_{ij}
S´	$n \times g$ reduced matrix of final scores
$\mathbf{e}_i$	$n \times 1$ unit vector with 1 in the ith element
$\mathbf{b_{h0k}}$	$n \times 1$ vector of seed scores for group k
$\mathbf{b_{hi}}$	$n \times 1$ vector of final affinity scores to node i
g	Number of groups (node partitions)
ϵ	$= 1/(1 + \max_i (d_{ii}))$ positive constant (< 1); encoding influence between neighbors
$\mathbf{DC_0}, \mathbf{DC}$	DELTACON$_0$, DELTACON
VEO	Vertex/Edge Overlap
GED	Graph Edit Distance [40]
SS	Signature Similarity [175]
λ-D Adj.	λ-distance on the Adjacency **A**
λ-D Lap	λ-distance on the Laplacian **L**
λ-D N.L.	λ-distance on the normalized Laplacian $\mathbf{L}_{norm} = \mathbf{D}^{-1/2} \mathbf{L} \mathbf{D}^{-1/2}$

Problem 5.1 DeltaConnectivity   Given (a) two graphs, $G_1(\mathcal{V}, \mathcal{E}_1)$ and $G_2(\mathcal{V}, \mathcal{E}_2)$ with the same node set,[1] $\mathcal{V}$, but different edge sets $\mathcal{E}_1$ and $\mathcal{E}_2$, and (b) the node correspondence. Find a similarity score, $sim(G_1, G_2) \in [0, 1]$, between the input graphs. A score of value 0 means totally different graphs, while 1 means identical graphs.

The obvious way to solve this problem is by measuring the overlap of their edges. Why does this often not work in practice? Consider the following example. According to the overlap method, the pairs of barbell graphs shown in Figure 5.2 of p. 116, $(B10, mB10)$ and $(B10, mmB10)$, have the same similarity score. But, clearly, from the aspect of information flow, a missing edge from a clique ($mB10$) does not play as important role in the graph connectivity as the missing "bridge" in $mmB10$. So, could we instead measure the differences in the 1-step away neighborhoods, 2-

---

[1]If the graphs have different, but overlapping node sets, $\mathcal{V}_1$ and $\mathcal{V}_2$, we assume that $\mathcal{V} = \mathcal{V}_1 \cup \mathcal{V}_2$, and the extra nodes are treated as singletons.

step away neighborhoods etc.? If yes, with what weight? DELTACON does that in a principled way (Observation 5.4, p. 100).

5.1.1   OVERVIEW

The first conceptual step of DELTACON is to compute the pairwise node affinities in the first graph, and compare them with the ones in the second graph. For notational compactness, we store them in a $n \times n$ similarity matrix[2] **S**. The s_{ij} entry of the matrix indicates the influence node i has on node j. For example, in a who-knows-whom network, if node i is, say, republican and if we assume homophily (i.e., neighbors are similar), how likely is it that node j is also republican? Intuitively, node i has more influence/affinity to node j if there are many short, heavily weighted paths from node i to j.

The second conceptual step is to measure the differences in the corresponding node affinity scores of the two graphs and report the result as their similarity score.

5.1.2   MEASURING NODE AFFINITIES

Pagerank [35], personalized Random Walks with Restarts (RWR) [93], lazy RWR [10], and the "electrical network analogy" technique [57] are only a few of the methods that compute node affinities. We could have used Personalized RWR: $[\mathbf{I} - (1 - c)\mathbf{A}\mathbf{D}^{-1}]\mathbf{b}_{\mathrm{h}i} = c\ \mathbf{e}_i$, where c is the probability of restarting the random walk from the initial node, $\mathbf{e}_i$ the starting (seed) indicator vector (all zeros except 1 at position i), and $\mathbf{b}_{\mathrm{h}i}$ the unknown Personalized Pagerank column vector. Specifically, s_{ij} is the affinity of node j with respect to node i. For reasons that we explain next, we chose to use Fast Belief Propagation (FABP [125]), an inference method that we introduced in Chapter 3. Specifically, we use a simplified form of FABP given in the following lemma:

Lemma 5.2   FABP *(Equation (3.4)) can be simplified and written as:*

$$[\mathbf{I} + \epsilon^2\mathbf{D} - \epsilon\mathbf{A}] \cdot \mathbf{b}_{\mathrm{h}i} = \mathbf{e}_i, \tag{5.1}$$

where $\mathbf{b}_{\mathrm{h}i} = [s_{i1}, \ldots s_{in}]^T$ *is the column vector of final similarity/influence scores starting from the* i^{th} *node,* ϵ *is a small constant capturing the influence between neighboring nodes,* **I** *is the identity matrix,* **A** *is the adjacency matrix and* **D** *is the diagonal matrix with the degree of node i as the* d_{ii} *entry.*

Proof. **(From FaBP to DeltaCon.)** We start from the equation for FABP in Chapter 3

$$[\mathbf{I} + a\mathbf{D} - c'\mathbf{A}]\mathbf{b}_{\mathrm{h}} = \boldsymbol{\phi}_h, \tag{3.4}$$

where $\boldsymbol{\phi}_h$ is the vector of prior scores, $\mathbf{b}_{\mathrm{h}}$ is the vector of final scores (beliefs), $a = 4h_h^2/(1 - 4h_h^2)$, and $c' = 2h_h/(1 - 4h_h^2)$ are small constants, and h_h is a small constant that encodes the influence between neighboring nodes (homophily factor). By using the Maclaurin approximation for

---

[2]In reality, we don't measure all the affinities (see Section 5.2.2 for an efficient approximation).

division in Table 3.4, we obtain

$$1/\left(1 - 4h_h^2\right) \approx 1 + 4h_h^2.$$

To obtain Equation (5.1), the core formula of DELTACON, we substitute the latter approximation in Equation (3.4), and also set $\boldsymbol{\phi_h} = \mathbf{e}_i$, $\mathbf{b_h} = \mathbf{b}_{\mathrm{h}i}$, and $h_h = \epsilon/2$. □

For an equivalent, more compact notation, we use the matrix form, and stack all the $\mathbf{b}_{\mathrm{h}i}$ vectors ($i = 1, \dots, n$) into the $n \times n$ matrix $\mathbf{S}$. We can easily prove that

$$\boxed{\mathbf{S} = [s_{ij}] = [\mathbf{I} + \epsilon^2 \mathbf{D} - \epsilon \mathbf{A}]^{-1}}. \tag{5.2}$$

Equivalence to Personalized RWR. Before we move on with the intuition behind our method, we note that the version of FABP that we use (Equation (5.1)) is identical to Personalized RWR under specific conditions, as shown in Theorem 5.3.

Theorem 5.3   *The FABP equation (Equation (5.1)) can be written in a Personalized RWR-like form:*

$$[\mathbf{I} - (1 - c'')\mathbf{A}_* \mathbf{D}^{-1}]\mathbf{b}_{\mathrm{h}i} = c'' \, \mathbf{y},$$

where $c'' = 1 - \epsilon$, $\mathbf{y} = \mathbf{A}_* \mathbf{D}^{-1} \mathbf{A}^{-1} \frac{1}{c''} \mathbf{e}_i$ *and* $\mathbf{A}_* = \mathbf{D}(\mathbf{I} + \epsilon^2 \mathbf{D})^{-1} \mathbf{D}^{-1} \mathbf{A} \mathbf{D}$.

Proof. We begin from the derived FABP equation (Equation (5.1)) and do simple linear algebra operations:

$[\mathbf{I} + \epsilon^2 \mathbf{D} - \epsilon \mathbf{A}]\mathbf{b}_{\mathrm{h}i} = \mathbf{e}_i$ $\hfill (\times \, \mathbf{D}^{-1} \text{ from the left})$

$[\mathbf{D}^{-1} + \epsilon^2 \mathbf{I} - \epsilon \mathbf{D}^{-1}\mathbf{A}]\mathbf{b}_{\mathrm{h}i} = \mathbf{D}^{-1}\mathbf{e}_i$ $\hfill (\mathbf{F} = \mathbf{D}^{-1} + \epsilon^2 \mathbf{I})$

$[\mathbf{F} - \epsilon \mathbf{D}^{-1}\mathbf{A}]\mathbf{b}_{\mathrm{h}i} = \mathbf{D}^{-1}\mathbf{e}_i$ $\hfill (\times \, \mathbf{F}^{-1} \text{ from the left})$

$[\mathbf{I} - \epsilon \mathbf{F}^{-1}\mathbf{D}^{-1}\mathbf{A}]\mathbf{b}_{\mathrm{h}i} = \mathbf{F}^{-1}\mathbf{D}^{-1}\mathbf{e}_i$ $\hfill (\mathbf{A}_* = \mathbf{F}^{-1}\mathbf{D}^{-1}\mathbf{A}\mathbf{D})$

$[\mathbf{I} - \epsilon \mathbf{A}_*\mathbf{D}^{-1}]\mathbf{b}_{\mathrm{h}i} = (1 - \epsilon)\,(\mathbf{A}_*\mathbf{D}^{-1}\mathbf{A}^{-1}\frac{1}{1-\epsilon}\mathbf{e}_i)$ $\hfill \square$

5.1.3   LEVERAGING BELIEF PROPAGATION

The reasons we choose BP and its fast approximation with Equation (5.2) are: (a) it is based on sound theoretical background (maximum likelihood estimation on marginals), (b) it is fast (linear on the number of edges), and (c) it agrees with intuition, taking into account not only direct neighbors, but also 2-, 3-, and k-step-away neighbors, with decreasing weight. We elaborate on the last reason, as follows.

Intuition 5.4   Attenuating Neighboring Influence
By temporarily ignoring the echo cancellation term $\epsilon^2 \mathbf{D}$ in Equation (5.2), we can expand the matrix inversion and approximate the $n \times n$ matrix of pairwise affinities, $\mathbf{S}$, as

$$\mathbf{S} \approx [\mathbf{I} - \epsilon \mathbf{A}]^{-1} \approx \mathbf{I} + \epsilon \mathbf{A} + \epsilon^2 \mathbf{A}^2 + \dots.$$

As we said, our method captures the differences in the 1-step, 2-step, 3-step, etc. neighborhoods in a weighted way; differences in long paths have a smaller effect on the computation of the similarity measure than differences in short paths. Recall that $\epsilon < 1$, and that $\mathbf{A}^k$ has information about the k-step paths. This intuition is related to the principle behind Katz centrality [111]. We note that this is just the intuition behind our method; we do not use this simplified formula to find matrix $\mathbf{S}$.

5.1.4   DESIRED PROPERTIES FOR SIMILARITY MEASURES

Let $G_1(\mathcal{V}, \mathcal{E}_1)$ and $G_2(\mathcal{V}, \mathcal{E}_2)$ be two graphs, and $sim(G_1, G_2) \in [0, 1]$ denote their similarity score. Then, we want the similarity measure to obey the following axioms.

- A1. *Identity property*: $sim(G_1, G_1) = 1$.

- A2. *Symmetric property*: $sim(G_1, G_2) = sim(G_2, G_1)$.

- A3. *Zero property*: $sim(G_1, G_2) \to 0$ for $n \to \infty$, where G_1 is the complete graph (K_n), and G_2 is the empty graph (i.e., the edge sets are complementary).

Moreover, the measure must be as follows.

(a) intuitive   It should satisfy the following desired properties.

P1. [*Edge Importance*] For unweighted graphs, changes that create disconnected components should be penalized more than changes that maintain the connectivity properties of the graphs.

P2. [*Edge-"Submodularity"*] For unweighted graphs, a specific change is more important in a graph with few edges than in a much denser, but equally sized graph.

P3. [*Weight Awareness*] In weighted graphs, the bigger the weight of the removed edge is, the greater the impact on the similarity measure should be.

In Section 5.2.3 we formalize the properties and discuss their satisfiability by our proposed similarity measure theoretically. Moreover, in Section 5.4 we introduce and discuss an additional, *informal*, property:

IP. [*Focus Awareness*] "Random" changes in graphs are less important than "targeted" changes of the same extent.

(b) scalable   The huge size of the generated graphs, as well as their abundance require a similarity measure that is computed fast and handles graphs with billions of nodes.

5.2   DELTACON: "δ" CONNECTIVITY CHANGE DETECTION

Now that we have described the high level ideas behind our method, we move on to the details.

5.2.1   ALGORITHM DESCRIPTION

Let the graphs we compare be $G_1(\mathcal{V}, \mathcal{E}_1)$ and $G_2(\mathcal{V}, \mathcal{E}_2)$. If the graphs have different node sets, say $\mathcal{V}_1$ and $\mathcal{V}_2$, we assume that $\mathcal{V} = \mathcal{V}_1 \cup \mathcal{V}_2$, where some nodes are disconnected.

As mentioned before, the main idea behind our proposed similarity algorithm is to compare the node affinities in the given graphs. The steps of our similarity method are as follows.

Step 1   By Equation (5.2), we compute for each graph the $n \times n$ matrix of pairwise node affinity scores ($\mathbf{S}_1$ and $\mathbf{S}_2$ for graphs G_1 and G_2, respectively).

Step 2   Among the various distance and similarity measures (e.g., Euclidean distance (ED), cosine similarity, correlation) found in the literature, we use the root Euclidean distance (RootED, a.k.a. Matusita distance)

$$d = \mathrm{RootED}(\mathbf{S}_1, \mathbf{S}_2) = \sqrt{\sum_{i=1}^{n}\sum_{j=1}^{n}(\sqrt{s_{1,ij}} - \sqrt{s_{2,ij}})^2}. \tag{5.3}$$

We use the RootED distance for the following reasons:

1. it is very similar to the Euclidean distance (ED), the only difference being the square root of the pairwise similarities (s_{ij});

2. it usually gives better results, because it "boosts" the node affinities[3] and, therefore, detects even small changes in the graphs (other distance measures, including ED, suffer from high similarity scores no matter how much the graphs differ); and

3. satisfies the desired properties $P1-P3$, as well as the informal property IP. As discussed in Section 5.2.3, at least $P1$ is not satisfied by the ED.

Step 3   For interpretability, we convert the distance (d) to a similarity measure (sim) via the formula $sim = \frac{1}{1+d}$. The result is bounded to the interval [0,1], as opposed to being unbounded [0,∞). Notice that the distance-to-similarity transformation does *not* change the ranking of results in a nearest-neighbor query.

The straightforward algorithm, DeltaCon$_0$ (Algorithm 5.1), is to compute all the n^2 affinity scores of matrix $\mathbf{S}$ by simply using Equation (5.2). We can do the inversion using the Power Method or any other efficient method.

---

[3]The node affinities are in [0, 1], so the square root makes them bigger.

Algorithm 5.1 DELTACON$_0$

Input : edge files of $G_1(\mathcal{V}, \mathcal{E}_1)$ and $G_2(\mathcal{V}, \mathcal{E}_2)$
Output : $sim(G_1, G_2)$

1: $\mathbf{S}_1 = [\mathbf{I} + \epsilon^2 \mathbf{D}_1 - \epsilon \mathbf{A}_1]^{-1}$ // $s_{1,ij}$: affinity/influence of
2: $\mathbf{S}_2 = [\mathbf{I} + \epsilon^2 \mathbf{D}_2 - \epsilon \mathbf{A}_2]^{-1}$ //node i to node j in G_1
3: $d(G_1, G_2) =$ ROOTED $(\mathbf{S}_1, \mathbf{S}_2)$
4: $sim(G_1, G_2) = \frac{1}{1+d(G_1,G_2)}$

5.2.2 FASTER COMPUTATION

DELTACON$_0$ satisfies all the properties in Section 5.1, but it is quadratic (n^2 affinity scores s_{ij} are computed by using the power method for the inversion of a sparse matrix) and thus not scalable. We present a faster, linear algorithm, DELTACON (Algorithm 5.2), which approximates DELTACON$_0$ and differs in the first step. We still want each node to become a seed exactly once in order to find the affinities of the rest of the nodes to it; but here we have multiple seeds at once, instead of having one seed at a time. The idea is to randomly divide our node-set into g groups, and compute the affinity score of each node i to group k, $\mathbf{S}'_k$, by solving the linear system $[\mathbf{I} + \epsilon^2 \mathbf{D} - \epsilon \mathbf{A}]\mathbf{S}'_k = \sum_{i \in group_k} \mathbf{e}_i$. This requires computing only $n \times g$ scores, which are then stored column-wise in the $n \times g$ matrix $\mathbf{S}'$ ($g \ll n$). Intuitively, instead of using the $n \times n$ affinity matrix $\mathbf{S}$, we add up the scores of the columns that correspond to the nodes of a group, and obtain the $n \times g$ matrix $\mathbf{S}'$ ($g \ll n$). The score s'_{ik} is the affinity of node i to the kth group of nodes ($k = 1, \ldots, g$). The following lemma gives the complexity of computing the node-group affinities.

Lemma 5.5 *The time complexity of computing the reduced affinity matrix, $\mathbf{S}'$, is linear on the number of edges.*

Proof. We can compute the $n \times g$ "skinny" matrix $\mathbf{S}'$ quickly, by solving $[\mathbf{I} + \epsilon^2 \mathbf{D} - \epsilon \mathbf{A}]\mathbf{S}' = [\mathbf{b}_{h01} \ldots \mathbf{b}_{h0g}]$, where $\mathbf{b}_{h0k} = \sum_{i \in group_k} \mathbf{e}_i$ is the membership $n \times 1$ vector for group k (all 0s, except 1s for members of the group). Solving this system is equivalent to solving for each group $k \in \{1, \ldots, g\}$ the linear system $[\mathbf{I} + \epsilon^2 \mathbf{D} - \epsilon \mathbf{A}]\mathbf{S}'_k = \mathbf{b}_{h0k}$ to obtain the node-to-group affinity vector $\mathbf{S}'_k$. Using the power method (Section 3.2.2), the linear system can be solved in time linear on the number of non-zeros of the matrix $\epsilon \mathbf{A} - \epsilon^2 \mathbf{D}$, which is equivalent to the number of edges, m, of the input graph G. Thus, the g linear systems require $\mathrm{O}(g \cdot m)$ time, which is still linear on the number of edges for a small constant g. It is worth noting that the g linear systems can be solved in parallel, since there are no dependencies, and then the overall time is simply $\mathrm{O}(m)$. $\square$

Thus, we compute g final scores per node, which denote its affinity to every *group* of seeds, instead of every seed node that we had in Equation (5.2). With careful implementation, DELTACON is linear on the number of edges and groups g. As we show in Section 5.4.3, it takes ~ 160 sec, on commodity hardware, for a 1.6-million-node graph. Once we have the reduced affinity matrices $\mathbf{S}'_1$ and $\mathbf{S}'_2$ of the two graphs, we use the ROOTED, to find the similarity between the $n \times g$ matrices of final scores, where $g \ll n$. The pseudocode of the DELTACON is given in Algorithm 5.2.

Algorithm 5.2 DELTACON

Input : edge files of $G_1(\mathcal{V}, \mathcal{E}_1)$ and $G_2(\mathcal{V}, \mathcal{E}_2)$ and g (groups: # of node partitions)
Output : $sim(G_1, G_2)$

1: $\{\mathcal{V}_j\}_{j=1}^g = $ random_partition($\mathcal{V}, g$) // g groups
2: // estimate affinity vector of nodes $i = 1, \dots, n$ to group k
3: **for** $k = 1 \to g$ **do**
4:    $\boldsymbol{\phi}_{\boldsymbol{h}k} = \sum_{i \in \mathcal{V}_k} e_i$
5:    solve $[\mathbf{I} + \epsilon^2 \mathbf{D}_1 - \epsilon \mathbf{A}_1]\mathbf{b}_{\mathbf{h}'_{1k}} = \boldsymbol{\phi}_{\boldsymbol{h}k}$
6:    solve $[\mathbf{I} + \epsilon^2 \mathbf{D}_2 - \epsilon \mathbf{A}_2]\mathbf{b}_{\mathbf{h}'_{2k}} = \boldsymbol{\phi}_{\boldsymbol{h}k}$
7: **end for**
8: $\mathbf{S}'_1 = [\mathbf{b}_{\mathbf{h}'_{11}} \; \mathbf{b}_{\mathbf{h}'_{12}} \; \dots \; \mathbf{b}_{\mathbf{h}'_{1g}}]; \; \mathbf{S}'_2 = [\mathbf{b}_{\mathbf{h}'_{21}} \; \mathbf{b}_{\mathbf{h}'_{22}} \; \dots \; \mathbf{b}_{\mathbf{h}'_{2g}}]$
9: // compare affinity matrices $\mathbf{S}'_1$ and $\mathbf{S}'_2$
10: $d(G_1, G_2) = $ ROOTED $(\mathbf{S}'_1, \mathbf{S}'_2)$
11: $sim(G_1, G_2) = \frac{1}{1+d(G_1,G_2)}$

In an attempt to see how our random node partitioning algorithm in the first step fares with respect to more principled partitioning techniques, we used METIS [108]. Essentially, such an approach finds the influence of *coherent* subgraphs to the rest of the nodes in the graph—instead of the influence of randomly chosen nodes to the latter. We found that the METIS-based variant of our similarity method gave intuitive results for most small, synthetic graphs, but not for the real graphs. This is probably related to the lack of good edge-cuts on sparse real graphs [134], and also the fact that changes within a group manifest less when a group consists of the nodes belonging to a single community than randomly assigned nodes.

Next we give the time complexity of DELTACON, as well as the relationship between the similarity scores of DELTACON$_0$ and DELTACON.

Lemma 5.6 *The time complexity of* DELTACON, *when applied in parallel to the input graphs, is linear on the number of edges in the graphs, i.e.,* $O(g \cdot \max\{m_1, m_2\})$.

Proof. By using the power method (Section 3.2.2), the complexity of solving Equation (5.1) is $O(m_i)$ for each graph ($i = 1, 2$). The node partitioning needs $O(n)$ time; the affinity algorithm is run g times in each graph, and the similarity score is computed in $O(gn)$ time. Therefore, the complexity of DELTACON is $O((g + 1)n + g(m_1 + m_2))$, where g is a small constant. Unless the graphs are trees, $|\mathcal{E}_i| > n$ and, thus, the complexity of the algorithm reduces to $O(g(m_1 + m_2))$. Assuming that the affinity algorithm is run on the graphs in parallel, since there is no dependency between the computations, DELTACON has complexity $O(g \cdot \max\{m_1, m_2\})$.   □

Before we give the relationship between the similarity scores computed by the two proposed methods, we introduce a helpful lemma.

Lemma 5.7   *The affinity score of each node to a group (computed by* DELTACON*) is equal to the sum of the affinity scores of the node to each one of the nodes in the group individually (computed by* DELTACON₀*).*

Proof.   Let $\mathbf{B} = \mathbf{I} + \epsilon^2 \mathbf{D} - \epsilon \mathbf{A}$. Then DELTACON₀ consists of solving for every node $i \in V$ the equation $\mathbf{B} \cdot \mathbf{b}_{\mathbf{h}i} = \mathbf{e}_i$; DELTACON solves the equation $\mathbf{B} \cdot \mathbf{b}'_{\mathbf{h}k} = \boldsymbol{\phi}_{\boldsymbol{h}k}$ for all groups $k \in (0, g]$, where $\boldsymbol{\phi}_{\boldsymbol{h}k} = \sum_{i \in group_k} \mathbf{e}_i$. Because of the linearity of matrix additions, it holds true that $\mathbf{b}'_{\mathbf{h}k} = \sum_{i \in group_k} \mathbf{b}_{\mathbf{h}i}$, for all groups k.   □

Theorem 5.8   DELTACON*'s similarity score between any two graphs* G_1, G_2 *upper bounds the actual* DELTACON₀*'s similarity score, i.e.,* $sim_{DC_0}(G_1, G_2) \leq sim_{DC}(G_1, G_2)$.

Proof.   Intuitively, grouping nodes blurs the influence information and makes the nodes seem more similar than originally.

More formally, let $\mathbf{S}_1, \mathbf{S}_2$ be the $n \times n$ final score matrices of G_1 and G_2 by applying DELTACON₀, and $\mathbf{S}'_1, \mathbf{S}'_2$ be the respective $n \times g$ final score matrices by applying DELTACON. We want to show that DELTACON₀'s distance

$$d_{DC_0} = \sqrt{\sum_{i=1}^{n} \sum_{j=1}^{n} \left(\sqrt{s_{1,ij}} - \sqrt{s_{2,ij}} \right)^2}$$

is greater than DELTACON's distance

$$d_{DC} = \sqrt{\sum_{k=1}^{g} \sum_{i=1}^{n} \left(\sqrt{s'_{1,ik}} - \sqrt{s'_{2,ik}} \right)^2}$$

or, equivalently, that $d_{DC_0}^2 > d_{DC}^2$. It is sufficient to show that for one group of DELTACON, the corresponding summands in d_{DC} are smaller than the summands in d_{DC_0} which are related to the nodes that belong to the group. By extracting the terms in the squared distances that refer to one group of DeltaCon and its member nodes in DeltaCon$_0$, and by applying Lemma 5.7, we obtain the following terms:

$$t_{DC_0} = \sum_{i=1}^{n} \sum_{j \in group} \left(\sqrt{s_{1,ij}} - \sqrt{s_{2,ij}} \right)^2$$

$$t_{DC} = \sum_{i=1}^{n} \left(\sqrt{\sum_{j \in group} s_{1,ij}} - \sqrt{\sum_{j \in group} s_{2,ij}} \right)^2.$$

Next, we concentrate again on a selection of summands (e.g., $i = 1$), we expand the squares and use the Cauchy-Schwartz inequality to show that

$$\sum_{j \in group} \sqrt{s_{1,ij} s_{2,ij}} < \sqrt{\sum_{j \in group} s_{1,ij} \sum_{j \in group} s_{2,ij}},$$

or, equivalently, that $t_{DC_0} > t_{DC}$.   □

5.2.3   DESIRED PROPERTIES

We have already presented our scalable algorithm for graph similarity in detail, and the only question that remains from a theoretic perspective is whether DeltaCon satisfies the axioms and properties presented in Section 5.1.4.

A1. *Identity Property*: $sim(G_1, G_1) = 1$.

The affinity scores, **S**, are identical for the input graphs, because the two linear systems in Algorithm 5.1 are exactly the same. Thus, the RootED distance d is 0, and the DeltaCon similarity, $sim = \frac{1}{1+d}$, is equal to 1.

A2. *Symmetric Property*: $sim(G_1, G_2) = sim(G_2, G_1)$.

Similarly, the equations that are used to compute $sim(G_1, G_2)$ and $sim(G_2, G_1)$ are the same. The only difference is the order of solving them in Algorithm 5.1. Therefore, both the RootED distance d and the DeltaCon similarity score sim are the same.

A3. *Zero Property*: $sim(G_1, G_2) \rightarrow 0$ for $n \rightarrow \infty$, where G_1 is the complete graph (K_n), and G_2 is the empty graph (i.e., the edge sets are complementary).

Proof. First, we show that all the nodes in a complete graph get final scores in $\{s_g, s_{ng}\}$, depending on whether they are included in group g or not. Then, it can be demonstrated that the scores have finite limits, and specifically $\{s_g, s_{ng}\} \rightarrow \{\frac{n}{2g} + 1, \frac{n}{2g}\}$ as $n \rightarrow \infty$ (for finite $\frac{n}{g}$). Given this condition, it can be derived that the RootED, $d(G_1, G_2)$, between the **S** matrices of the empty and the complete graph becomes arbitrarily large. So, $sim(G_1, G_2) = \frac{1}{1+d(G_1,G_2)} \rightarrow 0$ for $n \rightarrow \infty$.   □

DELTACON satisfies the three axioms that every similarity measure must obey. We elaborate on the satisfiability of the properties of $P1 - P3$ next.

P1. [Edge Importance] For unweighted graphs, changes that create disconnected components should be penalized more than changes that maintain the connectivity properties of the graphs.

Formalizing this property in its most general case with any type of disconnected components is hard, thus we focus on a well-understood and intuitive case: the barbell graph.

Proof. Assume that $\mathbf{A}$ is the adjacency matrix of an undirected barbell graph with two cliques of size n_1 and n_2, respectively (e.g., $B10$ with $n_1 = n_2 = 5$ in Figure 5.2) and (i_0, j_0) is the "bridge" edge. Without loss of generality we can assume that $\mathbf{A}$ has a block-diagonal form, with one edge (i_0, j_0) linking the two blocks:

$$a_{ij} = \begin{cases} 1 & \begin{array}{l} \text{for } i, j \in \{1, \ldots, n_1\} \text{ and } i \neq j \\ \text{or } i, j \in \{n_1 + 1, \ldots, n_2\} \text{ and } i \neq j \\ \text{or } (i, j) = (i_0, j_0) \text{ or } (i, j) = (j_0, i_0) \end{array} \\ 0 & \text{otherwise.} \end{cases}$$

Then, $\mathbf{B}$ is an adjacency matrix with elements

$$b_{ij} = \begin{cases} 0 & \text{for } \mathbf{one} \text{ pair } (i, j) \neq (i_0, j_0) \text{ and } (j_0, i_0) \\ a_{ij} & \text{otherwise,} \end{cases}$$

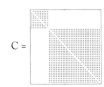

and $\mathbf{C}$ is an adjacency matrix with elements

$$c_{ij} = \begin{cases} 0 & \text{if } (i, j) = (i_0, j_0) \text{ or } (i, j) = (j_0, i_0) \\ a_{ij} & \text{otherwise.} \end{cases}$$

We want to prove that $sim(\mathbf{A}, \mathbf{B}) \geq sim(\mathbf{A}, \mathbf{C}) \Leftrightarrow d(\mathbf{A}, \mathbf{B}) \leq d(\mathbf{A}, \mathbf{C})$ or, equivalently, that $d^2(\mathbf{A}, \mathbf{B}) \leq d^2(\mathbf{A}, \mathbf{C})$.

From Equation (5.1), by expressing the matrix inversion using a power series and ignoring the terms of greater than second power, for matrix $\mathbf{A}$ we obtain the solution:

$$\mathbf{b}_{hi} = [\mathbf{I} + (\epsilon \mathbf{A} - \epsilon^2 \mathbf{D}_A) + (\epsilon \mathbf{A} - \epsilon^2 \mathbf{D}_A)^2 + \ldots]\mathbf{e}_i \Rightarrow \mathbf{S}_A = \mathbf{I} + \epsilon \mathbf{A} + \epsilon^2 \mathbf{A}^2 - \epsilon^2 \mathbf{D}_A,$$

where the index (e.g., $\mathbf{A}$) denotes the graph each matrix corresponds to. Now, using the last equation, we can write out the elements of the $\mathbf{S}_A, \mathbf{S}_B, \mathbf{S}_C$ matrices, and derive their ROOTED distances:

$$d^2(\mathbf{A}, \mathbf{B}) = 4(n_2 - f)\frac{\epsilon^4}{c_1^2} + 2\frac{\epsilon^2}{c_2^2}$$

$$d^2(\mathbf{A}, \mathbf{C}) = 2(n_1 + n_2 - 2)\epsilon^2 + 2\epsilon,$$

where $c_1 = \sqrt{\epsilon + \epsilon^2(n_2 - 3)} + \sqrt{\epsilon + \epsilon^2(n_2 - 2)}$ and $c_2 = \sqrt{\epsilon^2(n_2 - 2)} + \sqrt{\epsilon + \epsilon^2(n_2 - 2)}$, and $f = 3$ if the missing edge in graph B and the "bridge" edge are incident to the same node, or $f = 2$ in any other case.

We can, therefore write the difference of the distances we are interested in as

$$d^2(\mathbf{A}, \mathbf{C}) - d^2(\mathbf{A}, \mathbf{B}) = 2\epsilon \left(\epsilon(n_1 + n_2 - f) + 1 - \left(\frac{2\epsilon^3(n_1 - 2)}{c_1^2} + \frac{\epsilon}{c_2^2} \right) \right).$$

By observing that $c1 \geq 2\sqrt{(\epsilon)}$ and $c2 \geq \sqrt{\epsilon}$ for $n_2 \geq 3$ and using these facts in the last equation, it follows that:

$$d^2(\mathbf{A}, \mathbf{C}) - d^2(\mathbf{A}, \mathbf{B}) \geq 2\epsilon \left(\epsilon(n_1 + n_2 - f) - \frac{\epsilon^2(n_1 - 2)}{2} \right) \geq 2\epsilon^2 \left(n_1 + n_2 - f - \epsilon(n_1 - 2) \right).$$

Given that $n_2 \geq 3$ and $f = 2$ or 3, we obtain that $n_2 - f \geq 0$. Moreover, $0 < \epsilon < 1$ by definition and $n_1 - \epsilon n_1 \geq 0$. From these inequalities, it immediately follows that $d^2(\mathbf{A}, \mathbf{B}) \leq d^2(\mathbf{A}, \mathbf{C})$. We note that this property is not always satisfied by the Euclidean distance.   □

P2. [Edge-"Submodularity"] For unweighted graphs, a specific change is more important in a graph with few edges than in a much denser, but equally sized graph.

Intuition. Here, we provide simulation results that suggest that this property is satisfied by a simplified version of DELTACON (in the specific case of graph construction that we investigate). We leave a more in-depth exploration of its general form as future work.

Before presenting our empirical analysis, we formally define the property by introducing two pairs of graphs. Let $\mathbf{A}$ be the adjacency matrix of an undirected graph, with $m_\mathbf{A}$ non-zero elements (edges) a_{ij} and $a_{i_0 j_0} = 1$, and $\mathbf{B}$ be the adjacency matrix of another graph which is identical to $\mathbf{A}$, but is missing the edge (i_0, j_0).

$$b_{ij} = \begin{cases} 0 & \text{if } (i, j) = (i_0, j_0) \text{ or } (i, j) = (j_0, i_0) \\ a_{ij} & \text{otherwise} \end{cases}$$

Additionally, assume another pair of graphs with respective adjacency matrices $\mathbf{C}$ and $\mathbf{E}$[4] defined as follows.

$$c_{ij} = \begin{cases} 0 & \text{for } \geq 1 \text{ pair } (i, j) \neq (i_0, j_0) \text{ and } (i, j) \neq (j_0, i_0) \\ a_{ij} & \text{otherwise} \end{cases}$$

In other words, $\mathbf{C}$ has $m_\mathbf{C} < m_\mathbf{A}$ non-zero elements. Then, $\mathbf{E}$ is defined as the adjacency matrix of another graph which is identical to $\mathbf{C}$, but is missing the edge (i_0, j_0).

---

[4]We use $\mathbf{E}$ instead of $\mathbf{D}$ to distinguish the adjacency matrix of the graph from the diagonal matrix of degrees, which is normally defined as $\mathbf{D}$.

$$e_{ij} = \begin{cases} 0 & \text{for } (i,j) = (i_0, j_0) \text{ or } (i,j) = (j_0, i_0) \\ c_{ij} & \text{otherwise} \end{cases}$$

According to $P2$, we need to show that

$$sim(\mathbf{A}, \mathbf{B}) \geq sim(\mathbf{C}, \mathbf{E}) \Leftrightarrow d(\mathbf{A}, \mathbf{B}) \leq d(\mathbf{C}, \mathbf{E}).$$

By substituting the RootED distance, it turns out that we want to show that

$$\sqrt{\sum_{i=1}^{n}\sum_{j=1}^{n}(\sqrt{s_{\mathbf{A},ij}} - \sqrt{s_{\mathbf{B},ij}})^2} \leq \sqrt{\sum_{i=1}^{n}\sum_{j=1}^{n}(\sqrt{s_{\mathbf{C},ij}} - \sqrt{s_{\mathbf{E},ij}})^2},$$

where s_{ij} are the elements of the corresponding affinity matrix $\mathbf{S}$. These are defined for $\mathbf{A}$ by expressing the matrix inversion in Equation 5.1 with a power series, and ignoring the terms of greater than second power (simplified version or approximation of DeltaCon):

$$\mathbf{b}_{\mathbf{h}i} = [\mathbf{I} + (\epsilon\mathbf{A} - \epsilon^2\mathbf{D}_A) + (\epsilon\mathbf{A} - \epsilon^2\mathbf{D}_A)^2 + ...]\mathbf{e}_i \Rightarrow \mathbf{S}_\mathbf{A} = \mathbf{I} + \epsilon\mathbf{A} + \epsilon^2\mathbf{A}^2 - \epsilon^2\mathbf{D}_\mathbf{A},$$

where the index of $\mathbf{S}$ and $\mathbf{D}$ (e.g., $\mathbf{A}$) denotes the graph each matrix corresponds to.

For the empirical analysis of this property, we construct the graphs as follows: We start from a complete graph of size n, $G_0 = K_n$, and randomly pick an edge (i_0, j_0). Then, we generate a series of graphs, G_t, derived by G_{t-1} by removing one new edge (i_t, j_t). Note that (i_t, j_t) cannot be the initially chosen edge (i_0, j_0). For every derived graph, we compute the RootED distance between itself and the same graph without the edge (i_0, j_0). What we expect to see is that the distance decreases as the number of edges in the graph increases. In other words, the distance between a sparse graph and the same graph without (i_0, j_0) is bigger than the distance between a denser graph and the same graph missing the edge (i_0, j_0). The opposite holds for the DeltaCon similarity measure, but in the proofs we use distance since that function is mathematically easier to manipulate than the similarity function. In Figure 5.1, we plot, for different graph sizes n, the RootED distance as a function of the edges in the graph (from left to right the graph becomes denser, and tends to the complete graph K_n). For this graph construction, our simulations suggest that the submodularity property is satisfied by a simplified version of DeltaCon. Its analysis for more cases is of theoretical interest.   □

P3. [Weight Awareness] In weighted graphs, the bigger the weight of the removed edge is, the greater the impact on the similarity measure should be.

Lemma 5.9   *Assume that G_A is a graph with adjacency matrix $\mathbf{A}$ and elements $a_{ij} \geq 0$. Also, let G_B be a graph with adjacency matrix $\mathbf{B}$ and elements*

$$b_{ij} = \begin{cases} a_{ij} + k & \text{if } (i,j) = (i_0, j_0) \text{ or } (i,j) = (j_0, i_0) \\ a_{ij} & \text{otherwise} \end{cases}$$

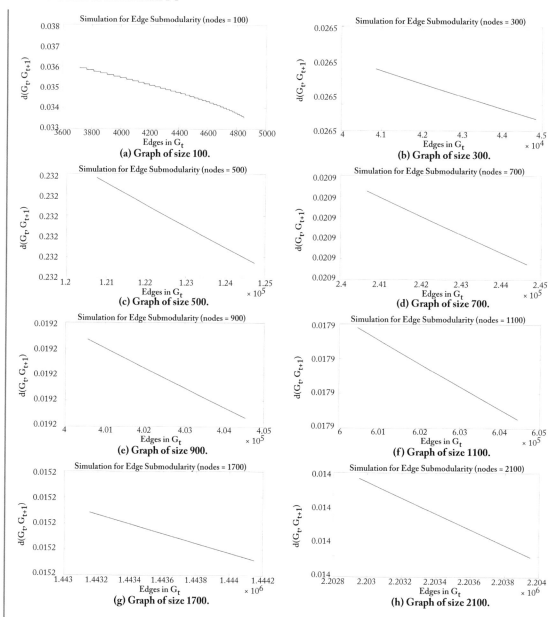

Figure 5.1: Illustration of submodularity via simulations for different graph sizes. In our simulations, the distance between two graphs that differ only in the edge (i_0, j_0) is a decreasing function of the number of edges in the original graph G_t (suggesting that the edge-submodularity property is satisfied). Reversely, their similarity is an increasing function of the number of edges in the original graph G_t.

where k is a positive integer (k ≥ 1). Then, it holds that $(S_B)_{ij} \geq (S_A)_{ij}$.

Proof. From Equation 5.1, by expressing the matrix inversion using a power series and ignoring the terms of greater than second power, we obtain the solution:

$$\mathbf{b}_{hi} = [\mathbf{I} + (\epsilon \mathbf{A} - \epsilon^2 \mathbf{D}) + (\epsilon \mathbf{A} - \epsilon^2 \mathbf{D})^2 + ...]\mathbf{e}_i \Rightarrow \mathbf{b}_{hi} \approx [\mathbf{I} + \epsilon \mathbf{A} + \epsilon^2 (\mathbf{A}^2 - \mathbf{D})]\mathbf{e}_i,$$

or equivalently

$$\mathbf{S}_A = \mathbf{I} + \epsilon \mathbf{A} + \epsilon^2 \mathbf{A}^2 - \epsilon^2 \mathbf{D}_A$$

and

$$\mathbf{S}_B = \mathbf{I} + \epsilon \mathbf{B} + \epsilon^2 \mathbf{B}^2 - \epsilon^2 \mathbf{D}_B.$$

We note that

$$\mathbf{S}_B - \mathbf{S}_A = \epsilon (\mathbf{B} - \mathbf{A}) + \epsilon^2 (\mathbf{B}^2 - \mathbf{A}^2) - \epsilon^2 (\mathbf{D}_B - \mathbf{D}_A), \tag{5.4}$$

where

$$(\mathbf{B} - \mathbf{A})_{ij} = \begin{cases} k & \text{if } (i, j) = (i_0, j_0) \text{ or } (i, j) = (j_0, i_0) \\ 0 & \text{otherwise} \end{cases}$$

and

$$(\mathbf{D}_B - \mathbf{D}_A)_{ij} = \begin{cases} k & \text{if } (i, j) = (i_0, j_0) \text{ or } (i, j) = (j_0, i_0) \\ 0 & \text{otherwise.} \end{cases}$$

By applying basic algebra, it can be shown that

$$(\mathbf{B}^2)_{ij} \begin{cases} = (\mathbf{A}^2)_{ij} + 2k \cdot a_{i_0 j_0} + k^2 & \text{if } (i, j) = (i_0, i_0) \text{ or } (i, j) = (j_0, j_0) \\ \geq (\mathbf{A}^2)_{ij} & \text{otherwise} \end{cases}$$

since $b_{ij} \geq a_{ij} \geq 0$ for all i, j by definition. Next, observe that for Equation 5.4, we have three cases:

- For all $(i, j) \neq (i_0, i_0)$ and (j_0, j_0), we have

$$(\mathbf{S}_B - \mathbf{S}_A)_{ij} = \epsilon (\mathbf{B} - \mathbf{A})_{ij} + \epsilon^2 (\mathbf{B}^2 - \mathbf{A}^2)_{ij} \geq 0$$

- For $(i, j) = (i_0, i_0)$ we have

$$(\mathbf{S}_B - \mathbf{S}_A)_{i_0 i_0} = \epsilon^2 k (2k \cdot a_{i_0 j_0} + k - 1) \geq 0$$

- For $(i, j) = (j_0, j_0)$ we have

$$(\mathbf{S}_B - \mathbf{S}_A)_{j_0 j_0} = \epsilon^2 k (2k \cdot a_{i_0 j_0} + k - 1) \geq 0$$

Hence, for all (i, j) it holds that $(S_B)_{ij} \geq (S_A)_{ij}$.   □

Next we will use the lemma to prove the weight awareness property.

Proof. [Property P3—Weight Awareness] We formalize the weight awareness property in the following way. Let $\mathbf{A}$ be the adjacency matrix of a weighted, undirected graph, with elements a_{ij}. Then, $\mathbf{B}$ is equal $\mathbf{A}$ but with a bigger weight for the edge (i_0, j_0), or more formally:

$$b_{ij} = \begin{cases} a_{ij} + k & \text{if } (i, j) = (i_0, j_0) \text{ or } (i, j) = (j_0, i_0) \\ a_{ij} & \text{otherwise.} \end{cases}$$

Let also $\mathbf{C}$ be the adjacency matrix of another graph with the same entries as $\mathbf{A}$ except for c_{i_0, j_0}, which is bigger than b_{i_0, j_0}:

$$c_{ij} = \begin{cases} a_{ij} + k' & \text{if } (i, j) = (i_0, j_0) \text{ or } (i, j) = (j_0, i_0) \\ a_{ij} & \text{otherwise,} \end{cases}$$

where $k' > k$ is an integer. To prove the property, it suffices to show that $sim(\mathbf{A}, \mathbf{B}) \geq sim(\mathbf{A}, \mathbf{C}) \Leftrightarrow d(\mathbf{A}, \mathbf{B}) \leq d(\mathbf{A}, \mathbf{C})$. Notice that this formal definition includes the case of removing an edge by assuming that $a_{i_0, j_0} = 0$ for matrix $\mathbf{A}$.

We can write the difference of the squares of the ROOTED distances as:

$$d^2(\mathbf{A}, \mathbf{B}) - d^2(\mathbf{A}, \mathbf{C}) = \sum_{i=1}^{n} \sum_{j=1}^{n} \left(\sqrt{s_{\mathbf{A}, ij}} - \sqrt{s_{\mathbf{B}, ij}} \right)^2 - \sum_{i=1}^{n} \sum_{j=1}^{n} \left(\sqrt{s_{\mathbf{A}, ij}} - \sqrt{s_{\mathbf{C}, ij}} \right)^2$$

$$= \sum_{i=1}^{n} \sum_{j=1}^{n} \left(\sqrt{s_{\mathbf{C}, ij}} - \sqrt{s_{\mathbf{B}, ij}} \right) \left(2\sqrt{s_{\mathbf{A}, ij}} - \sqrt{s_{\mathbf{B}, ij}} - \sqrt{s_{\mathbf{C}, ij}} \right) < 0$$

because $(\mathbf{S}_B)_{ij} \geq (\mathbf{S}_A)_{ij}$, $(\mathbf{S}_C)_{ij} \geq (\mathbf{S}_A)_{ij}$, and $(\mathbf{S}_C)_{ij} \geq (\mathbf{S}_B)_{ij}$ for all i, j by the construction of the matrices $\mathbf{A}$, $\mathbf{B}$, and $\mathbf{C}$, and Lemma 5.9. $\qquad\square$

In Section 5.4, we show experimentally that DELTACON not only satisfies the properties, but also that other similarity and distance methods fail in one or more test cases.

5.3   DELTACON-ATTR: ADDING NODE AND EDGE ATTRIBUTION

Thus far, we have broached the intuition and decisions behind developing a method for calculating graph similarity. However, we argue that computing this metric is only half the battle in the wider realm of change detection and graph understanding. Equally important is finding out *why* the graph changed the way it did. One way of doing this is attributing the changes to nodes and/or edges.

Equipped with this information, we can draw conclusions with respect to how certain changes impact graph connectivity and apply this understanding in a domain-specific context to assign blame, as well as instrument measures to prevent such changes in the future. Additionally, such a feature can be used to measure changes which have not yet happened in order to find information about which nodes and/or edges are most important for preserving or destroying connectivity. In this section, we will discuss our extension of a method, called DELTACON-ATTR, which enables node- and edge-level attribution for this very purpose.

5.3.1   ALGORITHM DESCRIPTION

Node Attribution   Our first goal is to find the nodes which are mostly responsible for the difference between the input graphs. Let the affinity matrices $\mathbf{S}'_1$ and $\mathbf{S}'_2$ be precomputed. Then, the steps of our node attribution algorithm (Algorithm 5.3) can be summarized to as follows.

Algorithm 5.3 DELTACON-ATTR Node Attribution

Input : affinity matrices $\mathbf{S}'_1$, $\mathbf{S}'_2$, edge files of $G_1(\mathcal{V}, \mathcal{E}_1)$ and $G_2(\mathcal{V}, \mathcal{E}_2)$, i.e., $\mathbf{A}_1$ and $\mathbf{A}_2$
Output : [$\mathbf{w}_{sorted}$, $\mathbf{w}_{sortedIndex}$] (node impact score sorted in descending order)

1: **for** $v = 1 \rightarrow n$ **do**
2:     // If an edge adjacent to the node has changed, the node is responsible:
3:     **if** $\sum |\mathbf{A}_1(v,:) - \mathbf{A}_2(v,:)| > 0$ **then**
4:        $w_v = \text{RootED}(\mathbf{S}'_{1,v}, \mathbf{S}'_{2,v})$
5:     **end if**
6: **end for**
7: // sort rows of vector w on column index 1
8: [$\mathbf{w}_{sorted}$, $\mathbf{w}_{sortedIndex}$] = sortRows($\mathbf{w}$, 1, 'descend')

Step 1   Intuitively, we compute the difference between the affinity of node v to the node groups in graph $\mathbf{A}_1$ and the affinity of node v to the node groups in graph $\mathbf{A}_2$. To that end, we use the same distance, RootED, that we applied to find the similarity between the whole graphs.

Given that the vth row vector ($v \leq n$) of $\mathbf{S}'_1$ and $\mathbf{S}'_2$ reflects the affinity of node v to the remainder of the graph, the RootED distance between the two vectors provides a measure of the extent to which that node is a *culprit* for change—we refer to this measure as the *impact* of a node. Thus, culprits with comparatively high impact are the ones that are the most responsible for change between graphs.

More formally, we quantify the contribution of each node to the graph changes by taking the RootED distance between each corresponding pair of row vectors in $\mathbf{S}'_1$ and $\mathbf{S}'_2$ as w_v for $v = 1, \ldots, n$ per Equation (5.5).

$$w_v = \text{RootED}(\mathbf{S}'_{1,v}, \mathbf{S}'_{2,v}) = \sqrt{\sum_{j=1}^{g} (\sqrt{s'_{1,vj}} - \sqrt{s'_{2,vj}})^2}. \tag{5.5}$$

Step 2   We sort the scores in the $n \times 1$ node impact vector w in descending order and report the most important scores and their corresponding nodes.

By default, we report culprits responsible for the top 80% of changes using a similar line of reasoning as that behind Fukunaga's heuristic [75]. In practice, we find that the notion of a skewed impact distribution holds (though the law of factor sparsity holds, the distribution need not be 80-20).

Edge Attribution   Complementarily to the node attribution approach, we have also developed an edge attribution method which ranks edge changes (additions and deletions) with respect to the graph changes. The steps of our edge attribution algorithm (Algorithm 5.4) are as follows.

Algorithm 5.4 DELTACON-ATTR Edge Attribution

Input : adjacency matrices $\mathbf{A}_1$, $\mathbf{A}_2$, culprit set of interest $\mathbf{w}_{sortedIndex,1 \ldots index}$ and node impact scores $\mathbf{w}$

Output : $\mathbf{E}_{sorted}$ (edge impact score) by descending value

---

1: **for** $v = 1 \rightarrow \text{length}(w_{sortedIndex,1 \ldots index})$ **do**
2:     srcNode = $w_{sortedIndex,v}$
3:     $\mathbf{r} = \mathbf{A}_{2,v} - \mathbf{A}_{1,v}$
4:     **for** $k = 1 \rightarrow n$ **do**
5:         destNode = k
6:         **if** $r_k = 1$ **then**
7:             edgeScore = $w_{\text{srcNode}} + w_{\text{destNode}}$
8:             append row [srcNode, destNode, '+', edgeScore] to $\mathbf{E}$
9:         **end if**
10:         **if** $r_k = -1$ **then**
11:             edgeScore = $w_{\text{srcNode}} + w_{\text{destNode}}$
12:             append row [srcNode, destNode, '-', edgeScore] to $\mathbf{E}$
13:         **end if**
14:     **end for**
15: **end for**
16: $\mathbf{E}_{sorted}$ = sortrows($\mathbf{E}$, 4, 'descend')   // sort rows of matrix $\mathbf{E}$ on column index 4

Step 1   We assign each changed edge incident to at least one node in the culprit set an impact score. This score is equal to the sum of impact scores for the nodes that the edge connects or disconnects.

Our goal here is to assign edge impact according to the degree that they affect the nodes they touch. Since even the removal or addition of a single edge does not necessarily impact both incident nodes equally, we choose the sum of both nodes' scores as the edge impact metric. Thus, our algorithm will rank edges which touch two nodes of moderately high impact more importantly than edges which touch one node of high impact but another of low impact.

Step 2   We sort the edge impact scores in descending order and report the edges in order of importance.

Analysis of changed edges can reveal important discrepancies from baseline behavior. Specifically, a large number of added edges or removed edges with individually low impact is indicative of star formation or destruction, whereas one or a few added or removed edges with individually high impact are indicative of community expansion or reduction via addition or removal of certain key bridge edges.

5.3.2   SCALABILITY ANALYSIS

Given precomputed S'_1 and S'_2 (precomputation is assumed since attribution can only be conducted after similarity computation), the node attribution component of DELTACON-ATTR is loglinear on the number of nodes, since n influence scores need to be sorted in descending order. In more detail, the cost of computing the impact scores for nodes is linear on the number of nodes and groups, but is dominated by the sorting cost given that $g \ll \log(n)$ in general.

With the same assumptions with respect to precomputed results, the edge attribution portion of DELTACON-ATTR is also loglinear, but on the sum of edge counts, since $m_1 + m_2$ total possible changed edges need to be sorted. In practice, the number of edges needing to be sorted should be far smaller, as we only need concern ourselves with edges which are incident to nodes in the culprit set of interest. Specifically, the cost of computing impact scores for edges is linear on the number of nodes in the culprit set k and the number of changed edges, but is again dominated by the sorting cost given that $k \ll \log(m_1 + m_2)$ in general.

5.4   EMPIRICAL RESULTS

We conduct several experiments on synthetic (Figure 5.2), as well as real data (Table 5.4 with undirected, unweighted graphs, unless stated otherwise) to answer the following questions.

> **Q1.** Does DELTACON agree with our intuition and satisfy the axioms/properties? Where do other methods fail?
> **Q2.** Is DELTACON scalable and able to compare large-scale graphs?
> **Q3.** How sensitive is it to the number of node groups?

We implemented the code in Matlab and ran the experiments on AMD Opteron Processor 854 @3 GHz, RAM 32 GB. The selection of parameters in Equation (5.1) follows the lines of Chapter 3—all the parameters are chosen so that the system converges.

5.4.1   INTUITIVENESS OF DELTACON

To answer Q1, for the first 3 properties (P1-P3), we conduct experiments on small graphs of 5–100 nodes and classic topologies (cliques, stars, circles, paths, barbell and wheel-barbell graphs, and "lollipops" shown in Figure 5.2), since people can argue about their similarities. For the name conventions, see Table 5.2. For our method we used five groups ($g = 5$), but the results are similar for other choices of the parameter. In addition to synthetic graphs, for informal property (IP), we use real networks with up to 11 million edges (Table 5.4).

We compare our method, DELTACON, to the six best state-of-the-art similarity measures that apply to our setting.

1. Vertex/Edge Overlap (VEO): In [167], the VEO similarity between two graphs $G_1(V_1, \mathcal{E}_1)$ and $G_2(V_2, \mathcal{E}_2)$ is defined as:

$$sim_{VEO}(G_1, G_2) = 2\frac{|\mathcal{E}_1 \cap \mathcal{E}_2| + |V_1 \cap V_2|}{|\mathcal{E}_1| + |\mathcal{E}_2| + |V_1| + |V_2|}.$$

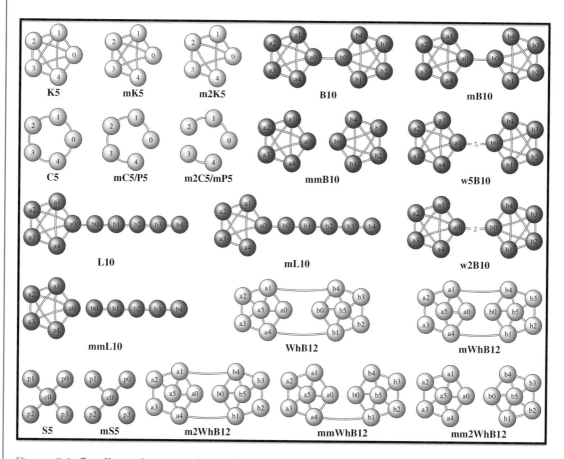

Figure 5.2: Small, synthetic graphs used in the DELTACON experimental analysis – *K: clique, C: cycle, P: path, S: star, B: barbell, L: lollipop, and WhB: wheel-barbell.* See Table 5.2 for the name conventions of the synthetic graphs.

Table 5.2: Name conventions for synthetic graphs. Missing number after the prefix implies $X = 1$.

Symbol	Meaning
K_n	Clique of size n
P_n	Path of size n
C_n	Cycle of size n
S_n	Star of size n
L_n	Lollipop of size n
B_n	Barbell of size n
WhB_n	Wheel barbell of size n
m_X	Missing X edges
mm_X	Missing X "bridge" edges
w	Weight of "bridge" edge

2. Graph Edit Distance (GED): GED has quadratic complexity in general, but it is linear on the number of nodes and edges when only insertions and deletions are allowed [38]:

$$sim_{GED}(G_1, G_2) \quad = \quad |\mathcal{V}_1| + |\mathcal{V}_2| - 2|\mathcal{V}_1 \cap \mathcal{V}_2| + |\mathcal{E}_1| + |\mathcal{E}_2| - 2|\mathcal{E}_1 \cap \mathcal{E}_2|.$$

For $\mathcal{V}_1 = \mathcal{V}_2$ and unweighted graphs, sim_{GED} is equivalent to the Hamming distance (HD) defined as $HD(\mathbf{A}_1, \mathbf{A}_2) = sum(\mathbf{A}_1 \; XOR \; \mathbf{A}_2)$.

3. Signature Similarity (SS): This is the best performing similarity measure studied in [167]. It starts from node and edge features, and by applying the SimHash algorithm (random projection based method), projects the features to a small dimension feature space, which is called *signature*. The similarity between the graphs is defined as the similarity between their signatures.

4. The last three methods are variations of the well-studied spectral method "λ-distance" ([38, 171, 222]). Let $\{\lambda_{1i}\}_{i=1}^{|\mathcal{V}_1|}$ and $\{\lambda_{2i}\}_{i=1}^{|\mathcal{V}_2|}$ be the eigenvalues of the matrices that represent G_1 and G_2. Then, λ-distance is given by

$$d_\lambda(G_1, G_2) = \sqrt{\sum_{i=1}^{k} (\lambda_{1i} - \lambda_{2i})^2},$$

where k is $\max(|\mathcal{V}_1|, |\mathcal{V}_2|)$ (padding is required for the smallest vector of eigenvalues). The variations of the method are based on three different matrix representations of the graphs: adjacency (λ-D Adj.), laplacian (λ-D Lap.) and normalized laplacian matrix (λ-D N.L.).

The results for the first three properties are presented in the form of Tables 5.3a–5.3b. For property P1 we compare the graphs (A,B) and (A,C) and report the difference between the pairwise similarities/distances of our proposed methods and the 6 state-of-the-art methods. We have arranged the pairs of graphs in such way that (A,B) are more similar than (A,C). Therefore, table entries that are non-positive mean that the corresponding method does not satisfy the property. Similarly, for properties P2 and P3, we compare the graphs (A,B) and (C,D) and report the difference in their pairwise similarity/distance scores.

P1. Edge Importance   *"Edges whose removal creates disconnected components are more important than other edges whose absence does not affect the graph connectivity. The more important an edge is, the more it should affect the similarity or distance measure."*

For this experiment we use the barbell, "wheel barbell" and "lollipop" graphs depicted in Figure 5.2, since it is easy to argue about the importance of the individual edges. The idea is that edges *in* a highly connected component (e.g., clique, wheel) are not very important from the information flow viewpoint, while edges that *connect* (almost uniquely) dense components play a significant role in the connectivity of the graph and the information flow. The importance of the "bridge" edge depends on the size of the components that it connects; the bigger the components, the more important the role of the edge is.

Observation 5.10   Only DELTACON succeeds in distinguishing the importance of the edges (*P1*) with respect to connectivity, while all the other methods fail at least once (Table 5.3a).

P2. "Edge-Submodularity"   *"Let $A(\mathcal{V}, \mathcal{E}_1)$ and $B(\mathcal{V}, \mathcal{E}_2)$ be two unweighted graphs with the same node set, and $|\mathcal{E}_1| > |\mathcal{E}_2|$ edges. Also, assume that $m_x A(\mathcal{V}, \mathcal{E}_1)$ and $m_x B(\mathcal{V}, \mathcal{E}_2)$ are the respective derived graphs after removing x edges. We expect that $sim(A, m_x A) > sim(B, m_x B)$, since the fewer the edges in a constant-sized graph, the more "important" they are."*

The results for different graph topologies and 1 or 10 removed edges (prefixes 'm' and 'm10' respectively) are given compactly in Table 5.3b. Recall that non-positive values denote violation of the "edge-submodularity" property.

Observation 5.11   Only DELTACON complies to the "edge-submodularity" property (*P2*) in all cases examined.

P3. Weight Awareness   *"The absence of an edge of big weight is more important than the absence of a smaller weighted edge; this should be reflected in the similarity measure."*

The weight of an edge defines the strength of the connection between two nodes, and, in this sense, can be viewed as a feature that relates to the importance of the edge in the graph. For

Table 5.3: DeltaCon₀ and DeltaCon (in bold) obey all the formal required properties (P1-P3). Each row of the tables corresponds to a comparison between the similarities (or distances) of two pairs of graphs; pairs (A,B) and (A,C) for property (P1); and pairs (A,B) and (C,D) for (P2) and (P3). Non-positive values of $\Delta s = sim(A, B) - sim(C, D)$ and $\Delta d = d(C, D) - d(A, B)$ for similarity and distance methods, respectively, are highlighted and mean violation of the property of interest.

(a) "Edge Important" (P1). Non-positive entries violate P1.

Graphs			DC_0	DC	VEO	SS	GED (XOR)	λ-d Adj.	λ-d Lap.	λ-d N.L.
A	B	C	$\Delta s = sim(A,B) - sim(A,C)$				$\Delta d = d(A,C) - d(A,B)$			
B10	mB10	mmB10	0.07	0.04	0	-10^{-5}	0	0.21	−0.27	2.14
L10	mL10	mmL10	0.04	0.02	0	10^{-5}	0	−0.30	−0.43	−8.23
WhB10	mWhB10	mmWhB10	0.03	0.01	0	-10^{-5}	0	0.22	0.18	−0.4.1
WhB10	m2WhB10	mm2WhB10	0.07	0.04	0	-10^{-5}	0	0.59	0.41	0.87

(b) "Edge Submodulatiry" (P2). Non-positive entries violate P2.

Graphs				DC_0	DC	VEO	SS	GED (XOR)	λ-d Adj.	λ-d Lap.	λ-d N.L.
A	B	C	D	$\Delta s = sim(A,B) - sim(C,D)$				$\Delta d = d(C,D) - d(A,B)$			
K5	mK5	C5	mC5	0.03	0.03	0.02	10^{-5}	0	−0.24	−0.59	−7.77
C5	mC5	P5	mP5	0.03	0.01	0.01	-10^{-5}	0	−0.55	−0.39	−0.20
K100	mK100	C100	mC100	0.03	0.02	0.002	10^{-5}	0	−1.16	−1.69	−311
C100	mC100	P100	mP100	10^{-4}	0.01	10^{-5}	-10^{-5}	0	−0.08	−0.06	−0.08
K100	m10K100	C100	m10C100	0.10	0.08	0.02	10^{-5}	0	−3.48	−4.52	−1089
C100	m10C100	P100	m10P100	0.001	0.001	10^{-5}	0	0	−0.03	0.01	0.31

(c) "Weight Awareness" (P3). Non-positive entries violate P3.

Graphs				DC_0	DC	VEO	SS	GED (XOR)	λ-d Adj.	λ-d Lap.	λ-d N.L.
A	B	C	D	$\Delta s = sim(A,B) - sim(C,D)$				$\Delta d = d(C,D) - d(A,B)$			
B10	mB10	B10	w5B10	0.09	0.08	−0.02	10^{-5}	−1	3.67	5.61	84.44
mmB10	B10	mm B10	w5B10	0.10	0.10	0	10^{-4}	0	4.57	7.60	95.61
B10	mB10	w5B10	w2B10	0.06	0.06	−0.02	10^{-5}	−1	2.55	3.77	66.71
w5B10	w2 B10	w5B10	mmB10	0.10	0.07	0.02	10^{-5}	1	2.23	3.55	31.04
w5B10	w2 B10	w5B10	B10	0.03	0.02	0	10^{-5}	0	1.12	1.84	17.73

Table 5.4: Large real and synthetic datasets

Name	Nodes	Edges	Description
Brain Graphs Small [170]	70	800–1,208	connectome
Enron Email [125]	36,692	367,662	who-emails-whom
Facebook wall [224]	45,813	183,412	who-posts-to-whom
Facebook links [224]	63,731	817,090	friend-to-friend
Epinions [94]	131,828	841,372	who-trusts-whom
Email EU [140]	265,214	420,045	who-sent-to-whom
Web Notre Dame [207]	325,729	1,497,134	site-to-site
Web Stanford [207]	281,903	2,312,497	site-to-site
Web Google [207]	875,714	5,105,039	site-to-site
Web Berk/Stan [207]	685,230	7,600,595	site-to-site
AS Skitter [140]	1,696,415	11,095,298	p2p links
Brain Graphs Big [170]	16,777,216	49,361,130–90,492,237	connectome
Kronecker 1	6,561	65,536	synthetic
Kronecker 2	19,683	262,144	synthetic
Kronecker 3	59,049	1,048,576	synthetic
Kronecker 4	177,147	4,194,304	synthetic
Kronecker 5	531,441	16,777,216	synthetic
Kronecker 6	1,594,323	67,108,864	synthetic

this property, we study the weighted versions of the barbell graph, where we assume that all the edges except the "bridge" have unit weight.

Observation 5.12   All the methods are weight-aware ($P3$), except VEO and GED, which compute just the overlap in edges and vertices between the graphs (Table 5.3c).

IP. Focus Awareness   At this point, all the competing methods have failed in satisfying at least one of the formal desired properties. To test whether DELTACON satisfies our *informal* property, i.e., it is able to distinguish the extent of a change in a graph, we analyze real datasets with up to 11 million edges (Table 5.4) for two different types of changes. For each graph we create corrupted instances by removing: (i) edges from the original graph randomly, and (ii) the same number of edges in a targeted way (we randomly choose nodes and remove all their edges, until we have removed the appropriate fraction of edges).

For this property, we study 8 real networks: Email EU and Enron Emails, Facebook wall and Facebook links, Google and Stanford web, Berkeley/Stanford web and AS Skitter. In Figure 5.3a–5.3d, we give the DELTACON similarity score between the original graph and the corrupted graph with up to 30% removed edges. For each graph, we perform the experiment for random (solid line) and targeted (dashed line) edge removal.

Observation 5.13

- *"Targeted changes hurt more."* DELTACON is focus-aware (IP). Removal of edges in a targeted way leads to smaller similarity of the derived graph to the original one than removal of the same number of edges in a random way.

- *"More changes: random $\approx$ targeted."* In Figure 5.3, as the fraction of removed edges increases, the similarity score for random changes (solid lines) tends to the similarity score for targeted changes (dashed lines).

This is expected, because the random and targeted edge removal tend to be equivalent when a significant fraction of edges is deleted.

In Figure 5.3e-5.3f, we give the similarity score as a function of the percent of the removed edges. Specifically, the x axis corresponds to the percentage of edges that have been removed from the original graph, and the y axis gives the similarity score. As before, each point maps to the similarity score between the original graph and the corresponding corrupted graph.

Observation 5.14 *"More changes hurt more."* The higher the corruption of the original graph, the smaller the DELTACON similarity between the derived and the original graph is. In Figure 5.3, we observe that as the percentage of removed edges increases, the similarity to the original graph decreases consistently for a variety of real graphs.

General Remarks. All in all, the baseline methods have several non-desirable properties. The spectral methods, as well as SS fail to comply with the "edge importance" (P1) and "edge submodularity" (P2) properties. Moreover, λ-distance has high computational cost when the whole graph spectrum is computed, cannot distinguish the differences between co-spectral graphs, and sometimes small changes lead to big differences in the graph spectra. VEO and GEDFocu are oblivious to significant structural properties of the graphs; thus, despite their straightforwardness and fast computation, they fail to discern various changes in the graphs. On the other hand, DELTACON gives tangible similarity scores and conforms to all the desired properties.

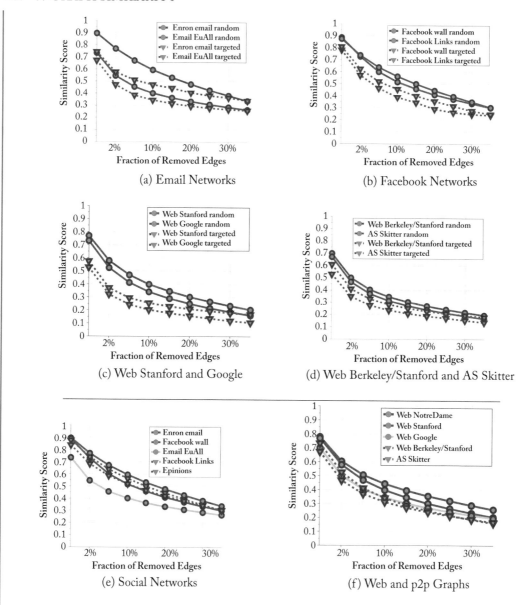

Figure 5.3: DELTACON is "focus-aware" (IP): Targeted changes hurt more than random ones. (a)–(d): DELTACON similarity scores for random (solid lines) and targeted (dashed lines) changes vs. the fraction of removed edges in the "corrupted" versions of the original graphs (x axis). We note that the dashed lines are always below the solid lines of the same color. (e)–(f): DELTACON agrees with intuition: the more a graph changes (i.e., the number of removed edges increases), the smaller is its similarity to the original graph.

5.4.2   INTUITIVENESS OF DELTACON-ATTR

In addition to evaluating the intuitiveness of DELTACON$_0$ and DELTACON, we also test DELTA-CON-ATTR on a number of synthetically created and modified graphs, and compare it to the state-of-the-art methods. We perform two types of experiments. The first experiment examines whether the *ranking* of the culprit nodes by our method agrees with intuition. In the second experiment, we evaluate DELTACON-ATTR's *classification* accuracy in finding culprits, and compare it to the best-performing competitive approach, CAD[5] [200], which was introduced concurrently, and independently from us. CAD uses the idea of commute time between nodes to define the anomalousness of nodes/edges. In a random walk, the commute time is defined as the expected number of steps starting at i, before node j is visited and then node i is reached again. We give a qualitative comparison between DELTACON-ATTR and CAD in Section 6.6 (Node/Edge Attribution).

Ranking Accuracy   We extensively tested DELTACON-ATTR on a number of synthetically created and modified graphs, and compared it with CAD. We note that CAD was designed to simply identify culprits in time-evolving graphs without ranking them. In order to compare it with our method, we adapted CAD so that it returns ranked lists of node and edge culprits: (i) We rank the culprit edges in decreasing order of edge score ΔE; (ii) To each node v, we attach a score equal to the sum of the scores of its adjacent edges, i.e., $\sum_{u \in N(v)} \Delta E((v, u))$, where $N(v)$ are the neighbors of v. Subsequently, we rank the nodes in decreasing order of attached score.

We give several of the conducted experiments in Table 5.5, and the corresponding graphs in Figure 5.4. Each row of the table corresponds to a comparison between graph A and graph B. The node and edge culprits that explain the main differences between the compared graphs are annotated in Figure 5.4. The darker a node is, the higher it is in the ranked list of node culprits. Similarly, edges that are adjacent to darker nodes are higher in the ranked list of edge culprits than edges that are adjacent to lighter nodes. If the returned ranked list agrees with the expected list (according to the formal and informal properties), we characterize the attribution of the method correct (checkmark). If there is disagreement, we provide the ordered list that the method returned. If two nodes or edges are tied, we use "=". For CAD we picked the parameter δ such that the algorithm returns 5 culprit edges and their adjacent nodes. Thus, we mark the returned list with "*" if CAD outputs 5 culprits while more exist. For each comparison, we also give the properties (last column) that define the order of the culprit edges and nodes.

Observation 5.15   DELTACON-ATTR reflects the desired properties (P1, P2, P3, and IP), while CAD fails to return the expected ranked lists of node and edge culprits in several cases.

Next we explain some of the comparisons that we present in Table 5.5.

---

[5]CAD was originally introduced for finding culprit nodes and edges without ranking them. We extended the proposed method to rank the culprits.

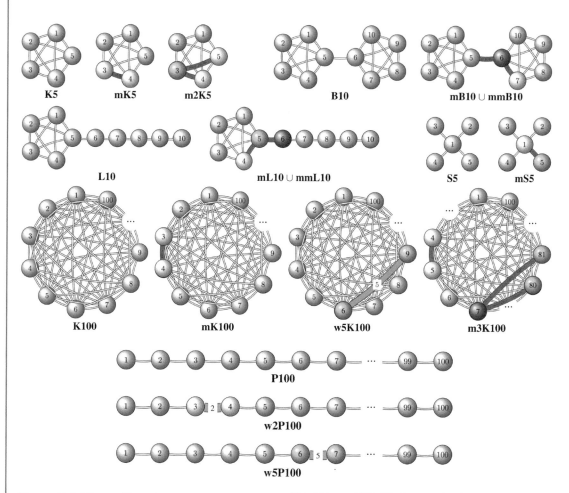

Figure 5.4: DELTACON-ATTR respects properties P1–P3, and IP. Nodes marked green are identified as the culprits for the change between the graphs. Darker shade corresponds to higher rank in the list of culprits. Removed and weighted edges are marked red and green, respectively. (*Continues.*)

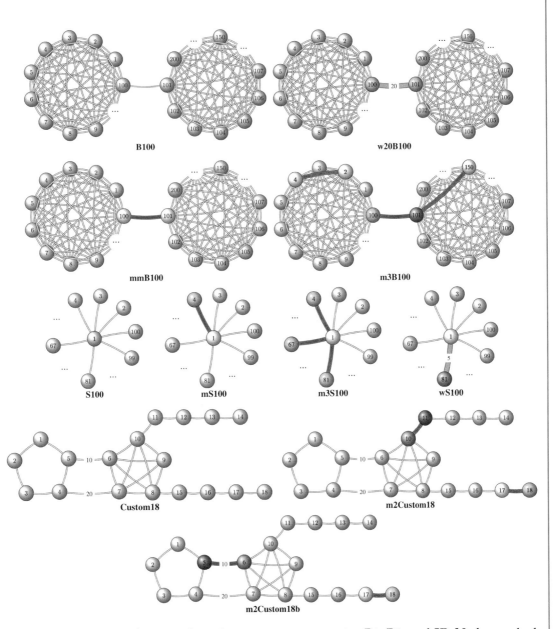

Figure 5.4: (*Continued.*) DELTACON-ATTR respects properties P1–P3, and IP. Nodes marked green are identified as the culprits for the change between the graphs. Darker shade corresponds to higher rank in the list of culprits. Removed edges are marked red.

Table 5.5: DELTACON-ATTR obeys all the required properties. Each row corresponds to a comparison between graph A and graph B, and evaluates the node and edge attribution of DELTACON-ATTR and CAD. The right order of edges and nodes is marked in Figures 5.4 and 5.4. We give the ranking of a method if it is different from the expected one.

Graphs		DeltaCon-Attr		CAD		Relevant
A	B	Edges	Nodes	Edges: δ for 1 = 5	Nodes	Properties
K5	mK5	✓	✓	✓	✓	
K5	m2K5	✓	✓	✓	✓	IP
B10	mB10 ∪ mmB10	✓	✓	✓	✓	P1, P2, IP
L10	mL10 ∪ mmL10	✓	✓	✓	5,6,4	P1, IP
S5	mS5	✓	✓	✓	1=5	P1, P2
K100	mK100	✓	✓	✓	✓	
K100	w5K100	✓	✓	✓	✓	P3
mK100	w5K100	✓	✓	✓	✓	P3,
K100	m3K100	✓	✓	✓	✓	P3, IP
K100	m10K100	✓	✓	(80,82)=(80,88)=(80,92)*	80,30,88=92*	P3, IP
P100	mP100	✓	✓	✓	✓	P1
w2P100	w5P100	✓	✓	✓	✓	P1, P3
B200	mmB200	✓	✓	✓	✓	P1
w20B200	m3B200	✓	✓	✓	✓	P1, P3, IP
S100	mS100	✓	✓	✓	1=4	P1, P2
S100	m3S100	✓	✓	✓	1,81=67=4	P1, P2, IP
wS100	m3S100	✓	✓	(1,4),(1,67),(1,81)	1,4=67,81	P1, P3, IP
Custom18	m2Custom18	✓	✓	(18,17),(10,11)	18,17,10,11	P1, P2
Custom18	m4Custom18b	✓	✓	✓	5=6,17=18	P1, P3

- **K5-mK5**: The pair consists of a 5-node complete graph and the same graph with one missing edge, $(3, 4)$. DELTACON-ATTR considers nodes 3 and 4 top culprits, with equal rank, due to equivalent loss in connectivity. Edge $(3, 4)$ is ranked top, and is essentially the only changed edge. CAD finds the same results.

- **K5-m2K5**: The pair consists of a 5-node complete graph and the same graph with two missing edges, $(3, 4)$ and $(3, 5)$. Both DELTACON-ATTR and CAD consider node 3 the top culprit, because two of its adjacent edges were removed. Node 3 is followed by 4 and

5, which are tied since they are both missing one adjacent edge (Property IP). The removed edges, $(3, 4)$ and $(3, 5)$, are considered equally responsible for the difference between the two input graphs. We observe similar behavior in larger complete graphs with 100 nodes (K100, and modified graphs mK100, w5K100, etc.). In the case of K100 and m10K100,[6] CAD does not return all 13 node culprits and 10 culprit edges because its parameter, δ, was set so that it would return at most 5 culprit edges.[7]

- **B10-mB10 $\cup$ mmB10**: We compare a barbell graph of 10 nodes to the same graph that is missing both an edge from a clique, $(6, 7)$, and the bridge edge, $(5, 6)$. As expected, DELTACON-ATTR finds 6, 5, and 7 as top culprits, where 6 is ranked higher than 5, since 6 lost connectivity to both nodes 5 and 7, whereas 5 disconnected only from 6. Node 5 is ranked higher than 7 because the removal of the bridge edge is more important than the removal of $(6, 7)$ within the second clique (Property P1). CAD returns the same results. We observe similar results in the case of the larger barbell graphs (B200, mmB200, w20B200, m3B200).

- **L10-mL10 $\cup$ mmL10**: This pair of graphs corresponds to the lollipop graph, $L10$, and the lollipop variant, $mL10 \cap mmL10$, that is missing one edge from the clique, as well as a bridge edge. Nodes 6, 5 and 4 are considered the top culprits for the difference in the graphs. Moreover, 6 is ranked more responsible for the change than 5, since 6 lost connectivity to a more strongly connected component than 5 (Property P2). However, CAD ranks node 5 higher than node 6 despite the difference in the connectivity of the two components (violation of P2).

- **S5, mS5**: We compare a 5-node star graph, and the same graph missing the edge $(1, 5)$. DELTACON-ATTR considers 5 and 1 top culprits, with 5 ranking higher than 1, as the edge removal caused a loss of connectivity from node 5 to all the peripheral nodes of the star, $2, 3, 4$, and the central node, 1. CAD considers nodes 1 and 5 equally responsible, ignoring the difference in the connectivity of the components (violation of P2). Similar results are observed in the comparisons between the larger star graphs—S100, mS100, m3S100, wS100.

- **Custom18-m2Custom18**: The ranking of node culprits that DELTACON-ATTR finds is 11, 10, 18, and 17. The nodes 11 and 10 are considered more important than the nodes 18 and 17, as the edge removal $(10, 11)$ creates a large connected component and a small chain of 4 nodes, while the edge removal $(17, 18)$ leads to a single isolated node (18). Node 10 is higher in the culprit list than node 11 because it loses connectivity to a denser component.

---

[6]m10K100 is a complete graph of 100 nodes where we have removed 10 edges: (i) 6 of the edges were adjacent to node 80—$(80, 82), (80, 84), (80, 86), (80, 88), (80, 90), (80, 92)$; (ii) 3 of the edges were adjacent to node 30—$(30, 50), (30, 60), (30, 70)$; and (iii) edge $(1, 4)$.

[7]The input graphs are symmetric. If edge (a, b) is considered culprit, CAD returns both (a, b) and (b, a).

The reasoning is similar for the ranking of nodes 18 and 17. CAD does not consider the differences in the density of the components, and leads to a different ranking of the nodes.

- **Custom18-m2Custom18**: The ranking of node culprits that DELTACON-ATTR returns is 5, 6, 18, and 17. This is in agreement with properties P1 and P3, since the edge $(5, 6)$ is more important than the edge $(17, 18)$. Node 5 is more responsible than node 6 for the difference between the two graphs, as node 5 ends up having reduced connectivity to a denser component. This property is ignored by CAD, which thus results in different node ranking.

As we observe, in all the synthetic and easily controlled examples, the ranking of the culprit nodes and edges that DELTACON-ATTR finds agrees with intuition.

Classification Accuracy   To further evaluate the accuracy of DELTACON-ATTR in classifying nodes as culprits, we perform a simulation-based experiment and compare our method to CAD. Specifically, we set up a simulation similar to the one that was introduced in [200].

We sample 2,000 points from a 2-dimensional Gaussian mixture distribution with four components, and construct the matrix $\mathbf{P} \in \mathcal{R}^{2000 \times 2000}$, with entries $p(i, j) = \exp \|i - j\|$, for each pair of points (i, j). Intuitively, the adjacency matrix $\mathbf{P}$ corresponds to a graph with four clusters that have strong connections within them, but weaker connections across them. By following the same process and adding some noise in each component of the mixture model, we also build a matrix $\mathbf{Q}$, and add more noise to it, which is defined as:

$$\mathbf{R}_{ij} = \begin{cases} 0 & \text{with probability } 0.95 \\ u_{ij} \sim \mathcal{U}(0, 1) & \text{otherwise,} \end{cases}$$

where $\mathcal{U}(0, 1)$ is the uniform distribution in $(0, 1)$. Then, we compare the two graphs, G_A and G_B, to each other, which have adjacency matrices $\mathbf{A} = \mathbf{P}$ and $\mathbf{B} = \mathbf{Q} + (\mathbf{R} + \mathbf{R}')/2$, respectively. We consider culprits (or anomalous) the inter-cluster edges for which $\mathbf{R}_{ij} \neq 0$, and the adjacent nodes. According to property P1, these edges are considered important (major culprits) for the difference between the graphs, as they establish more connections between loosely coupled clusters.

Conceptually, DELTACON-ATTR and CAD are similar because they are based on related methods [125] (Belief Propagation and Random Walk with Restarts, respectively). As shown in Figure 5.5, the simulation described above corroborates this argument, and the two methods have comparable performance—i.e., the areas under the ROC curves are similar for various realizations of the data described above. Over 15 trials, the AUC of DELTACON-ATTR and CAD is 0.9922 and 0.9934, respectively.

Observation 5.16   Both methods are very accurate in detecting nodes that are responsible for the differences between two highly-clustered graphs (Property P1).

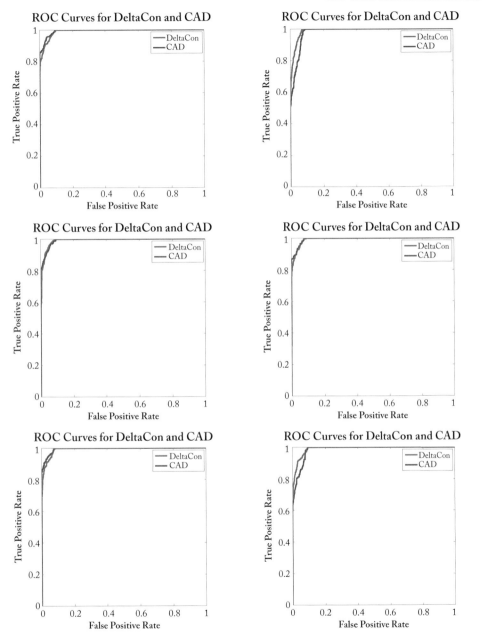

Figure 5.5: DELTACON-ATTR ties state-of-the-art method with respect to accuracy. Each plot shows the ROC curves for DELTACON-ATTR and CAD for different realizations of two synthetic graphs. The graphs are generated from points sampled from a 2-dimensional Gaussian mixture distribution with four components.

All in all, both DELTACON-ATTR and CAD have very high accuracy in detecting culprit nodes and edges that explain the differences between two input graphs. DELTACON-ATTR satisfies all the desired properties that define the importance of edges, while CAD sometimes fails to return the expected ranked lists of culprits.

5.4.3   SCALABILITY

In Section 5.1 we demonstrated that DELTACON is linear on the number of edges, and here we show that this also holds in practice. We ran DELTACON on Kronecker graphs (Table 5.4), which are known [131] to share many properties with real graphs.

Observation 5.17   As shown in Figure 5.6, DELTACON scales linearly with the number of edges in the largest input graph.

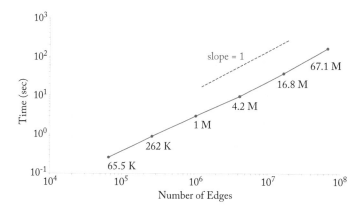

Figure 5.6: Scalability of DELTACON. DELTACON is linear on the number of edges (time in seconds vs. number of edges). The exact number of edges is annotated.

We note that the algorithm can be trivially parallelized by finding the node affinity scores of the two graphs in parallel instead of sequential. Moreover, for each graph, the computation of the similarity scores of the nodes to each of the g groups can be parallelized. However, the runtime of our experiments refer to the sequential implementation. The amount of time taken for DELTACON-ATTR is trivial even for large graphs, given that the necessary affinity matrices are already in memory from the DELTACON similarity computation. Specifically, node and edge attribution are log-linear on the nodes and edges, respectively, given that sorting is unavoidable for the task of ranking.

5.4.4   ROBUSTNESS

DELTACON$_0$ satisfies all the desired properties, but its runtime is quadratic and does not scale well to large graphs with more than a few million edges. On the other hand, our second proposed algorithm, DELTACON, is scalable both in theory and practice (Lemma 5.6, Section 5.4.3). In this section we present the sensitivity of DELTACON to the number of groups g, as well as how the similarity scores of DELTACON and DELTACON$_0$ compare.

For this experiment, we use complete and star graphs, as well as the Political Blogs dataset. For each of the synthetic graphs (a complete graph with 100 nodes and star graph with 100 nodes), we create three corrupted versions where we remove 1, 3, and 10 edges, respectively. For the real dataset, we create four corrupted versions of the Political Blogs graph by removing {10%, 20%, 40%, 80%} of the edges. For each pair of <original, corrupted> graphs, we compute the DELTACON similarity for varying number of groups. We note that when $g = n$, DELTACON is equivalent to DELTACON$_0$. The results for the synthetic and real graphs are shown in Figures 5.7a and 5.7b, respectively.

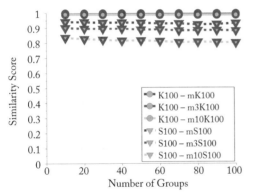

(a) Robustness of Method on Synthetic Graphs

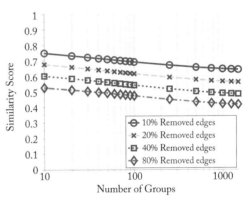

(b) Robustness of Method on the Political Blogs Dataset

Figure 5.7: DELTACON is robust to the number of groups. And more importantly, at every group level, the ordering of similarities of the different graph pairs remains the same (e.g., $sim(K100, m1K100) > sim(K100, m3K100) > \ldots > sim(S100, m1S100) > \ldots > sim(S100, m10S100)$).

Observation 5.18   In our experiments, DELTACON$_0$ and DELTACON agree on the ordering of pairwise graph similarities.

In Figure 5.7b the lines not only do not cross, but are almost parallel to each other. This means that for a fixed number of groups g, the differences between the similarities of the different graph pairs remain the same. Equivalently, the ordering of similarities is the same at every group level g.

Observation 5.19   The similarity scores of DELTACON are robust to the number of groups.

Obviously, the bigger the number of groups, the closer are the similarity scores to the "ideal" similarity scores, i.e., scores of DELTACON$_0$. For instance, in Figure 5.7b, when each blog is in its own group ($g = n = 1490$), the similarity scores between the original network and the derived networks (with any level of corruption) are identical to the scores of DELTACON$_0$. However, even with few groups, the approximation of DELTACON$_0$ is good.

It is worth mentioning that the bigger the number of groups g, the bigger runtime is required, since the complexity of the algorithm is $O(g \cdot max\{m1, m2\}|)$. Not only the accuracy, but also the runtime increases with the number of groups; so, the speed and accuracy trade-offs need to be conciliated. Experimentally, a good compromise is achieved even for g smaller than 100.

5.5   APPLICATIONS

In this section we present three applications of our graph similarity algorithms, one of which comes from social networks and the other two from the area of neuroscience.

5.5.1   ENRON

Graph Similarity   First, we employ DELTACON to analyze the ENRON dataset, which consists of emails sent among employees in a span of more than two years. Figure 5.8 depicts the DELTACON similarity scores between consecutive daily who-emailed-whom graphs. By applying Quality Control with Individual Moving Range, we obtain the lower and upper limits of the in-control similarity scores. These limits correspond to median $\pm 3\sigma$.[8] Using this method, we were able to define the threshold (lower control limit) below which the corresponding days are anomalous, i.e., they differ "too much" from the previous and following days. Note that all the anomalous days relate to crucial events in the company's history in 2001 (points marked with red boxes in Figure 5.8).

1. May 22, 2001: Jordan Mintz sends a memorandum to Jeffrey Skilling (CEO for a few months) for his sign-off on LJM paperwork;
2. August 21, 2001: Kenneth Lay, the CEO of Enron, emails all employees stating he wants "to restore investor confidence in Enron;"
3. September 26, 2001: Lay tells employees that the accounting practices are "legal and totally appropriate", and that the stock is "an incredible bargain;"

---

[8]The median is used instead of the mean, since appropriate hypothesis tests demonstrate that the data does not follow the normal distribution. Moving range mean is used to estimate σ.

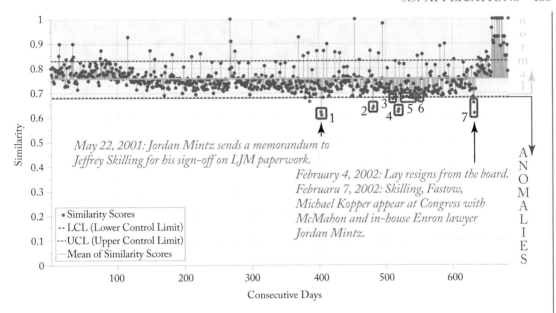

Figure 5.8: ENRON: Similarities between consecutive days and events. DELTACON detects anomalies on the Enron data coinciding with major events. The marked days correspond to anomalies. The blue points are similarity scores between consecutive instances of the daily email activity between the employees, and the marked days are 3σ units away from the median similarity score.

4. October 5, 2001: Just before Arthur Andersen hired Davis Polk & Wardwell law firm to prepare a defense for the company;

5. October 24–25, 2001: Jeff McMahon takes over as CFO. Email to all employees states that all the pertinent documents should be preserved;

6. November 8, 2001: Enron announces it overstated profits by 586 million dollars over five years;

7. February 4, 2002: Lay resigns from board.

Although high similarities between consecutive days do not consist anomalies, we found that mostly weekends expose high similarities. For instance, the first two points of 100% similarity correspond to the weekend before Christmas in 2000 and a weekend in July, when only two employees sent emails to each other. It is noticeable that after February 2002, many consecutive days are very similar; this happens because, after the collapse of Enron, the email exchange activity was rather low and often between certain employees.

Attribution We additionally apply DELTACON-ATTR to the ENRON dataset for the months of May 2001 and February 2002, which are the most anomalous months according to the analysis of the data on a month-to-month timescale. Based on the node and edge rankings produced as a result of our method, we drew some interesting real-world conclusions.

May 2001:

- Top Influential Culprit: John Lavorato, the former head of Enron's trading operations and CEO of Enron America, connected to ~50 new nodes in this month.

- Second Most Influential Culprit: Andy Zipper, VP of Enron Online, maintained contact with all those from the previous month, but also connected to 12 new people.

- Third Most Influential Culprit: Louise Kitchen, another employee (President of ENRON Online) lost 5-6 connections and made 5-6 connections. Most likely, some of the connections she lost or made were very significant in terms of expanding/reducing her office network.

February 2002:

- Top Influential Culprit: Liz Taylor lost 51 connections this month, but made no new ones—it is reasonable to assume that she likely quit the position or was fired (most influential culprit).

- Second-Most Influential Culprit: Louise Kitchen (third culprit in May 2001) made no new connections, but lost 22 existing ones.

- Third-Most Influential Culprit: Stan Horton (CEO of Enron Transportation) made 6 new connections and lost none. Some of these connections are likely significant in terms of expanding his office network.

- Fourth-, Fifth-, and Sixth-Most Influential Culprits: Employees Kam Keiser, Mike Grigsby (former VP for Enron's Energy Services), and Fletcher Sturm (VP) all lost many connections and made no new ones. Their situations were likely similar to those of Liz Taylor and Louise Kitchen.

5.5.2 BRAIN CONNECTIVITY GRAPH CLUSTERING

We also use DELTACON for the clustering and classification of graphs. For this purpose we study *connectomes*, i.e., brain graphs, which are obtained by Multimodal Magnetic Resonance Imaging [85].

In total, we study the connectomes of 114 subjects, which are related to attributes such as age, gender, IQ, etc. Each graph consists of 70 cortical regions (nodes), and connections (weighted edges) between them (see Table 5.4 "Brain Graphs Small"). We ignore the strength of connections and derive one undirected, unweighted brain graph per subject.

We first compute the DELTACON pairwise similarities between the brain graphs, and then perform hierarchical clustering using Ward's method (Figure 5.9b). As shown in the figure, there are two clearly separable groups of brain graphs. Applying a t-test on the available attributes for the two groups created by the clusters, we have found that the latter differ significantly ($p < .01$) in the Composite Creativity Index (CCI), which is related to the person's performance on a series of creativity tasks. Figure 5.9 illustrates the brain connections in a subject with high and low creativity index. It appears that more creative subjects have more and heavier connections across their hemispheres than those subjects that are less creative. Moreover, the two groups correspond to significantly different openness index ($p = .0558$), one of the "Big Five Factors;" that is, the brain connectivity is different in people that are inventive and people that are consistent. Exploiting analysis of variance (ANOVA: generalization of the t-test when more than two groups are analyzed), we tested whether the various clusters that we obtain from the connectivity-based hierarchical clustering map to differences in other attributes. However, in the dataset we studied there is no sufficient statistical evidence that age, gender, IQ, etc. are related to brain connectivity.

5.5.3   RECOVERY OF CONNECTOME CORRESPONDENCES

We also applied our method on the KKI-42 dataset [162, 185], which consists of connectomes corresponding to $n = 21$ subjects. Each subject underwent two Functional MRI scans at different times, and so the dataset has $2n = 42$ large connectomes with ~ 17 million voxels and 49.3–90.4 million connections among them (see Table 5.4 "Brain Graphs Big"). Our goal is to recover the pairs of connectomes that correspond to the same subject, by relying only on the structures of the brain graphs. In the following analysis, we compare our method to the standard approach in neuroscience literature [185], the Euclidean distance (as induced by the Frobenius norm), and also to the baseline approaches we introduced in Section 5.4.

We ran the following experiments on a 32-cores Intel(R) Xeon(R) CPU E7-8837 at 2.67 GHz, with 1 TB of RAM. The signature similarity method runs out of memory for these large graphs, and we could not use it for this application. Moreover, the variants of λ-distance are computationally expensive, even when we compute only a few top eigenvalues, and they also perform very poorly in this task.

Unweighted graphs. The brain graphs that were obtained from the FMRI scans have weighted edges that correspond to the strength of the connections between different voxels. The weights tend to be noisy, so we initially ignore them, and treat the brain graphs as binary by mapping all the non-zero weights to 1. To discover the pairs of connectomes that correspond to the same subject, we first find the DELTACON pairwise similarities between the connectomes. We note that it suffices to do $\binom{2n}{2} = 861$ graph comparisons, since the DELTACON similarities are symmetric. Then, we find the potential pairs of connectomes that belong to the same subject by using the following approach: For each connectome $C_i \in \{1, \ldots, 2n\}$, we choose the connectome $C_j \in \{1, \ldots, 2n\} \setminus i$ such that the similarity score, $sim(C_i, C_j)$, is maximized. In other words, we pair each connectome with its most similar graph (excluding itself) as defined by DELTACON. This results in 97.62% accuracy of predicting the connectomes of the same subject.

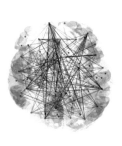

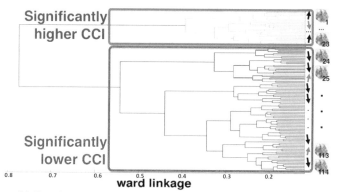

Significantly
higher CCI

Significantly
lower CCI

ward linkage

(a) Connectome: neural
network of brain.

(b) Dendogram representing the hierarchical clustering of the
DeltaCon similarities between the 114 connectomes.

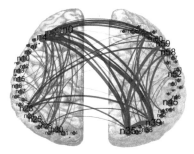

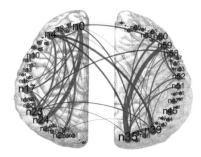

(c) Brain graph of a subject with
high creativity index.

(c) Brain graph of a subject with
low creativity index.

Figure 5.9: DELTACON-based clustering shows that artistic brains seem to have different wiring than the rest. (a) Brain network (connectome). Different colors correspond to each of the 70 cortical regions, whose centers are depicted by vertices. (b) Hierarchical clustering using the DELTACON similarities results in two clusters of connectomes. Elements in red correspond to mostly high creativity score. (c)–(d) Brain graphs for subjects with high and low creativity index, respectively. The low-CCI brain has fewer and lighter cross-hemisphere connections than the high-CCI brain.

In addition to our method, we compute the pairwise *Euclidean distances* (ED) between the connectomes and evaluate the predictive power of ED. Specifically, we compute the quantities $ED(i, j) = ||C_i - C_j||_F^2$, where C_i and C_j are the binary adjacency matrices of the connectomes i and j, respectively, and $|| \cdot ||_F$ is the Frobenius norm of the enclosed matrix. As before, for each connectome $i \in \{1, \ldots, 2n\}$, we choose the connectome $j \in \{1, \ldots, 2n\} \setminus i$ such that $ED(i, j)$

is minimized,[9] the accuracy of recovering the pairs of connectomes that correspond to the same subject is 92.86% (vs. 97.62% for DeltaCon), as shown in Figure 5.10.

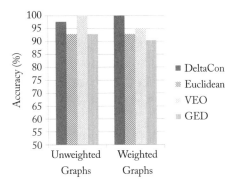

Figure 5.10: DeltaCon outperforms almost all the baseline approaches. It recovers correctly *all* pairs of connectomes that correspond to the same subject (100% accuracy) for *weighted* graphs outperforming all the baselines. It also recovers almost all the pairs correctly in the case of *unweighted* graphs, following VEO, which has the best accuracy.

Finally, from the baseline approaches, the Vertex/Edge Overlap similarity performs slightly better than our method, while GED has the same accuracy as the Euclidean distance (Figure 5.10, "Unweighted Graphs"). All the variants of λ-distance perform very poorly with 2.38% accuracy. As mentioned before, the signature similarity runs out of memory and, hence, could not be used for this application.

Weighted graphs. We also wanted to see how accurately the methods can recover the pairs of *weighted* connectomes that belong to the same subject. Given that the weights are noisy, we follow the common practice and first smooth them by applying the logarithm function (with base 10). Then we follow the procedure described above both for DeltaCon and the Euclidean distance. Our method yields 100% accuracy in recovering the pairs, while the Euclidean distance results in 92.86% accuracy. The results are shown in Figure 5.10 (labeled "Weighted Graphs"), while Figure 5.11 shows a case of brain graphs that were incorrectly recovered by the ED-based method, but successfully found by DeltaCon.

In the case of weighted graphs, as shown in Figure 5.10, all the baseline approaches perform worse than DeltaCon in recovering the correct brain graph pairs. In the plot we include the methods that have comparable performance to our method. The λ-distance has the same very poor accuracy (2.38% for all the variants) as in the case of unweighted graphs, while the signature similarity could not be applied due to its very high memory requirements.

---

[9]We note that DeltaCon computes the similarity between two graphs, while ED computes their distance. Thus, when trying to find the pairs of connectomes that belong to the same subject, we want to maximize the similarity, or equivalently, minimize the distance.

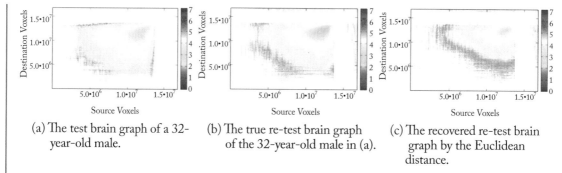

(a) The test brain graph of a 32-year-old male.

(b) The true re-test brain graph of the 32-year-old male in (a).

(c) The recovered re-test brain graph by the Euclidean distance.

Figure 5.11: DELTACON outperforms the Euclidean distance in recovering the correct test-retest pairs of brain graphs. We depict the spy plots of three brain graphs, across which the order of nodes is the same and corresponds to increasing order of degrees for the leftmost spy plot. The correct test-retest pair (a)–(b) that DELTACON recovers consists of "visually" more similar brain graphs than the incorrect test-retest pair (a)–(c) that the Euclidean distance found.

Therefore, by using DELTACON we are able to recover, with almost perfect accuracy, which large connectomes belong to the same subject. On the other hand, the commonly used technique, the Euclidean distance, as well as the baseline methods fail to detect several connectome pairs (with the exception of VEO in the case of unweighted graphs).

5.6 RELATED WORK

The related work comprises three main areas: Graph similarity, node affinity algorithms, and anomaly detection with node/edge attribution. We give the related work in each area separately, and mention what sets our method apart.

Graph Similarity Graph similarity refers to the problem of quantifying how similar two graphs are. The graphs may have the same or different sets of nodes, and can be divided into two main categories:

(1) *With Known Node Correspondence.* The first category assumes that the two given graphs are aligned, or, in other words, the node correspondence between the two graphs is given. [167] proposes five similarity measures for directed web graphs, which are applied for anomaly detection. Among them the best is Signature Similarity (SS), which is based on the SimHash algorithm, while Vertex/Edge Overlap similarity (VEO) also performs very well. Bunke [38] presents techniques used to track sudden changes in communications networks for performance monitoring. The best approaches are the Graph Edit Distance and Maximum Common Subgraph. Both are NP-complete, but the former approach can be simplified given the application and it becomes linear on the number of nodes and edges in the graphs. This chapter attacks

the graph similarity problem with known node correspondence, and is an extension of the work in [128], where DELTACON was first introduced. In addition to computational methods to assess the similarity between graphs, there is also a line of work on visualization-based graph comparison. These techniques are based on the side-by-side visualization of the two networks [15, 92], or superimposed/augmented graph or matrix views [12, 61]. A review of visualization-based comparison of information based on these and additional techniques is given in [80]. [12] investigate ways of visualizing differences between small brain graphs using either augmented graph representations or augmented adjacency matrices. However, their approach works for small and sparse graphs (40-80 nodes). Honeycomb [213] is a matrix-based visualization tool that handles larger graphs with several thousands edges, and performs temporal analysis of a graph by showing the time series of graph properties that are of interest. The visualization methods do not compute the similarity score between two graphs, but only show their differences. This is related to the culprit nodes and edges that our method DELTACON finds. However, these methods tend to visualize all the differences between two graphs, while our algorithm routes attention to the nodes and edges that are mostly responsible for the differences among the input graphs. All in all, visualizing and comparing graphs with millions or billions of nodes and edges remains a challenge, and best suits small problems.

(2) *With Unknown Node Correspondence.* The previous works assume that the correspondence of nodes across the two graphs is known, but this is not always the case. Social network analysis, bioinformatics, and pattern recognition are just a few domains with applications where the node correspondence information is missing or even constitutes the objective. The works attacking this problem can be divided into three main approaches: (a) feature extraction and similarity computation based on the feature space, (b) graph matching and the application of techniques from the first category, and (c) graph kernels.

There are numerous works that follow the first approach and use features to define the similarity between graphs. The λ-distance, a spectral method which defines the distance between two graphs as the distance between their spectra (eigenvalues) has been studied thoroughly ([38, 171, 222], algebraic connectivity [69]). The existence of co-spectral graphs with different structure, and the big differences in the graph spectra, despite subtle changes in the graphs, are two weaknesses that add up to the high complexity of computing the whole graph spectrum. Clearly, the spectral methods that call for the whole spectrum cannot scale to the large-scale graphs with billions of nodes and edges that are of interest currently. Also, depending on the graph-related matrix that is considered (adjacency, laplacian, normalized laplacian), the distance between the graphs is different. As we show in Section 6.4, these methods fail to satisfy one or more of the desired properties for graph comparison. [137] proposes an SVM-based approach on some global feature vectors (including the average degree, eccentricity, number of nodes and edges, number of eigenvalues, and more) of the graphs in order to perform graph classification. Macindoe and Richards [145] focus on social networks and extract three socially relevant features: leadership, bonding, and diversity. The complexity of the last feature makes the method applicable to graphs with up to a few million edges. Other techniques include computing edge curvatures under heat kernel embedding [59], comparing the number of spanning trees [112], comparing graph signatures consisting of summarized local structural features [25], and a distance based on graphlet correlation [227].

The second approach first solves the graph matching or alignment problem—i.e., finds the "best" correspondence between the nodes of the two graphs—and then finds the distance (or similarity) between the graphs. [49] reviews graph matching algorithms in pattern recognition. There are over 150 publications that attempt to solve the graph alignment problem under different settings and constraints. The methods span from genetic algorithms to decision trees, clustering, expectation-maximization and more. Some recent methods that are more efficient for large graphs include a distributed, belief-propagation-based method for protein alignment [34], another message-passing algorithm for aligning sparse networks when some possible matchings are given [23], and a gradient-descent-based method for aligning probabilistically large bipartite graphs (Chapter 6, [127]).

The third approach uses kernels between graphs, which were introduced in 2010 by [214]. Graph kernels work directly on the graphs without doing feature extraction. They compare graph structures, such as walks [77, 110], paths [32], cycles [95], trees [146, 180], and graphlets [51, 192] which can be computed in polynomial time. A popular graph kernel is the random walk graph kernel [77, 110], which finds the number of common walks on the two input graphs. The simple version of this kernel is slow, requiring $O(n^6)$ runtime, but can be sped up to $O(n^3)$ by using the Sylvester equation. In general, the above-mentioned graph kernels do not scale well to graphs with more than 100 nodes. A faster implementation of the random walk graph kernel with $O(n^2)$ runtime was proposed by Kang et al. [105]. The fastest kernel to date is the subtree kernel proposed by Shervashidze and Borgwardt [190, 191], which is linear on the number of edges and the maximum degree, $O(m \cdot d)$, in the graphs. The proposed approach uses the Weisfeiler-Lehman test of isomorphism, and operates on *labeled* graphs. In our work, we consider large, unlabeled graphs, while most kernels require at least $O(n^3)$ runtime or labels on the nodes/edges. Thus, we do not compare them to DELTACON quantitatively.

Remark. Both research problems—graph similarity with given or missing node correspondence—are important, but apply in different settings. If the node correspondence is available, the algorithms that make use of it can only perform better than the methods that omit it. The methods discussed in this chapter tackle the former problem. A guide to selecting a network similarity method is presented in [199].

Node Affinity   There are numerous node affinity algorithms; Pagerank [35], Personalized Random Walks with Restarts [93], the electric network analogy [57], SimRank [98] and extensions/improvements [230], [136], and Belief Propagation [228] are only some examples of the most successful techniques. In this chapter we focus on the latter method, and specifically a fast variation that we introduced in Chapter 3 ([125]). All the techniques have been success-fully in many tasks, such as ranking, classification, malware and fraud detection ([43],[150]), and recommendation systems [114].

Node/Edge Attribution   Detection of anomalous behaviors in time-evolving networks is more relevant to our work, and is covered in the surveys [8, 181]. Some anomaly detection methods discover anomalous nodes, and other anomalous structures in the graphs. In a slightly different context, a number of techniques have been proposed in the context of node and edge importance in graphs. PageRank, HITS [118] and betweenness centrality (random-walk-based [159] and shortest-path-based [73]) are several such methods for the purpose of identifying important nodes. [210] proposes a method to determine edge importance for the purpose of augmenting or inhibiting dissemination of information between nodes. To the best of our knowledge, this and other existing methods focus only on identifying important nodes and edges in the context of a single graph. In the context of anomaly detection, [6] and [200] detect nodes that contribute mostly to change events in time-evolving networks.

Among these works, the most relevant to ours are the methods proposed by [6] and [200]. The former relies on the selection of features, and tends to return a large number of false positives. Moreover, because of the focus on local egonet features, it may not distinguish between small and large changes in time-evolving networks [200]. At the same time, and independently from us, Sricharan and Das proposed CAD [200], a method which defines the anomalousness of edges based on the commute time between nodes. The commute time is the expected number of steps in a random walk starting at i, before node j is visited and then node i is reached again. This method is related to DELTACON as Belief Propagation (the heart of our method), and Random Walks with Restarts (the heart of CAD) are equivalent under certain conditions [125]. However, the methods work in different directions: DELTACON first identifies the most anomalous nodes, and then defines the anomalousness of edges as a function of the outlierness of the adjacent nodes; CAD first identifies the most anomalous edges, and then defines all their adjacent nodes as anomalous without ranking them. DELTACON does not only find anomalous nodes and edges in a graph, but also (i) ranks them in decreasing order of anomalousness (useful for guiding attention to important changes) and (ii) quantifies the difference between two graphs (which can also be used for graph classification, clustering, and other tasks).

CHAPTER 6

Graph Alignment

Can we spot the *same* people in two different social networks, such as LinkedIn and Facebook? How can we find *similar* people across different graphs? How can we effectively link an information network with a social network to support cross-network search? In all these settings, a key step is to align[1] the two graphs in order to reveal similarities between the nodes of the two networks. While in the previous chapter we focused on computing the similarity between two aligned networks, in this chapter we focus on aligning the nodes of two graphs, when that information is missing. Informally, the problem we tackle is defined as follows.

Problem 6.1   Graph Alignment or Matching—Informal Given two graphs, $G_1(\mathcal{V}_\infty, \mathcal{E}_\infty)$ and $G_2(\mathcal{V}_\in, \mathcal{E}_\in)$ where $\mathcal{V}$ and $\mathcal{E}$ are their node and edge sets, respectively. Find how to permute their nodes, so that the graphs have as similar structure as possible.

Graph alignment is a core building block in many disciplines, as it essentially enables us to link different networks so that we can search and/or transfer valuable knowledge across them. The notions of graph similarity and alignment appear in many disciplines such as protein-protein alignment [22, 34], chemical compound comparison [197], information extraction for finding synonyms in a single language, or translation between different languages [23], similarity query answering in databases [151], and pattern recognition [49, 232].

In this chapter we primarily focus on the alignment of *bipartite* graphs, i.e., graphs whose edges connect two disjoint sets of vertices (that is, there are no edges within the two node sets). Bipartite graphs stand for an important class of real graphs and appear in many different settings, such as author-conference publishing graphs, user-group membership graphs, and user-movie rating graphs. Despite their ubiquity, most (if not all) of the existing work on graph alignment is tailored for unipartite graphs and, thus, might be sub-optimal for bipartite graphs. In Table 6.1, we give the major symbols and definitions that we use in this chapter.

---

[1]Throughout this work we use the words "align(ment)" and "match(ing)" interchangeably.

Table 6.1: Description of major symbols

Symbol	Description
$\mathbf{P}$	user-level (node-level) correspondence matrix
$\mathbf{Q}$	group-level (community-level) correspondence matrix
$\mathbf{P}^{(v)}$	row or column vector of matrix P
$\mathbf{1}$	vector of 1s
$||A||_F$	$= \sqrt{Tr(A^T A)}$, Frobenius norm of $\mathbf{A}_1$
λ, μ	sparsity penalty parameters for $\mathbf{P}, \mathbf{Q}$, respectively (equivalent to lasso regularization)
η_P, η_Q	step of gradient descent for $\mathbf{P}, \mathbf{Q}$
ϵ	small constant (> 0) for the convergence of gradient descent

6.1 PROBLEM FORMULATION

In the past three decades, numerous communities have studied the problem of graph alignment, as it arises in many settings. However, most of the research has focused on *unipartite* graphs, i.e., graphs that consist of only one type of nodes. Formally, the problem that has been addressed in the past [56, 212, 216, 232] is the following: Given two *unipartite* graphs, G_1 and G_2, with adjacency matrices $\mathbf{A}_1$ and $\mathbf{A}_2$, find the permutation matrix $\mathbf{P}$ that minimizes the cost function f_{uni}:

$$\min_{\mathbf{P}} f_{uni}(\mathbf{P}) = \min_{\mathbf{P}} ||\mathbf{P A}_1 \mathbf{P}^T - \mathbf{A}_2||_F^2,$$

where $|| \bullet ||_F$ is the Frobenius norm of the corresponding matrix. We list the frequently used symbols in Table 6.1. The permutation matrix $\mathbf{P}$ is a square binary matrix with exactly one entry 1 in each row and column, and 0s elsewhere. Effectively, it reorders the rows of the adjacency matrix $\mathbf{A}_1$, while its transpose reorders the columns of the matrix, so that the resulting reordered matrix is "close" to $\mathbf{A}_2$.

In this work, we introduce the problem of aligning *bipartite* graphs. One example of such graphs is the user-group graph; the first set of nodes consists of users, the second set of groups, and the edges represent user memberships. Throughout the chapter we will consider the alignment of the "user-group" LinkedIn graph ($\mathbf{A}_1$) with the "user-group" Facebook graph ($\mathbf{A}_2$). In a more general setting, the reader may think of the first set consisting of nodes, and the second set of communities. First, we extend the traditional *unipartite* graph alignment problem definition to *bipartite* graphs.

Problem 6.2 Adaptation of Traditional Definition Given two **bipartite** graphs, G_1 and G_2, with adjacency matrices $\mathbf{A}_1$ and $\mathbf{A}_2$, find the **permutation** matrices $\mathbf{P}$ and $\mathbf{Q}$ that minimize the cost function f_0:

$$\min_{P,Q} f_0(P,Q) = \min_{P,Q} ||PA_1Q - A_2||_F^2,$$

where $|| \bullet ||_F$ is the Frobenius norm of the matrix.

This formulation uses two different permutation matrices to reorder the rows and columns of A_1 "independently," but it has two main shortcomings.

[S1] It is hard to solve, due to its *combinatorial* nature.

[S2] The permutation matrices imply that we are in search for *hard assignments* between the nodes of the input graphs. However, finding hard assignments might not be possible nor realistic. For instance, in the case of input graphs with a perfect "star" structure, aligning their spokes (peripheral nodes) is impossible, as they are identical from the structural viewpoint. In other words, any way of aligning the spokes is *equiprobable*. In similar (potentially more complicated) cases, soft assignment may be more valuable than hard assignment.

To deal with these issues, we relax Problem 6.2 that is directly adapted from the well-studied case of unipartite graphs [56, 212, 216, 232], and state it more realistically.

Problem 6.3 Soft, Sparse, Bipartite Graph Alignment Given two **bipartite** graphs, G_1 and G_2, with adjacency matrices A_1 and A_2, find the **correspondence** matrices P, Q that minimize the cost function f:

$$\min_{P,Q} f(P,Q) = \min_{P,Q} ||PA_1Q - A_2||_F^2$$

under the following constraints:

(1) [Probabilistic] Each matrix element is a probability, i.e., $0 \leq P_{ij} \leq 1$ and $0 \leq Q_{ij} \leq 1$, and

(2) [Sparsity] The matrices are sparse, i.e., $||P^{(v)}||_0 \leq t$ and $||Q^{(v)}||_0 \leq t$ for a small constant $t > 0$. The $|| \bullet ||_0$ denotes the l_0-"norm," i.e., the number of non-0s.

The *first constraint* of non-integer entries for the matrices has two advantages.

[A1] It solves both shortcomings of the traditional-based problem. The optimization problem is easier to solve, and has a realistic, probabilistic interpretation; it does not provide only the 1-to-1 correspondences, but also reveals the similarities between the nodes across networks. The entries of the correspondence matrix P (or Q) describe the probability that a LinkedIn user (or group) corresponds to a Facebook user (or group). We note that these properties are not guaranteed when the correspondence matrix is required to be a permutation or even doubly stochastic (square matrix with non-negative real entries, where each row and column sums to 1), which is common practice in the literature.

[A2] The matrices $\mathbf{P}$ and $\mathbf{Q}$ do *not* have to be square, which means that the matrices $\mathbf{A}_1$ and $\mathbf{A}_2$ can be of different size. This is yet another realistic requirement, as very rarely do two networks have the same number of nodes. Therefore, our formulation addresses not only graph alignment, but also *subgraph* alignment.

The *second constraint* follows naturally from the first one, as well as the large size of the social, and other networks. We want the correspondence matrices to be as sparse as possible, so that they encode few potential correspondences per node. Allowing every user/group of LinkedIn to be matched to every user/group of Facebook is not realistic and, actually, it is problematic for large graphs, as it has quadratic space cost with respect to the size of the input graphs.

To summarize, the existing approaches do not distinguish the nodes by types (e.g., users and groups), treat the graphs as unipartite, and, thus, aim at finding a permutation matrix $\mathbf{P}$, which gives a hard assignment between the nodes of the input graphs. In contrast, our formulation separates the nodes in categories, and can find correspondences at different granularities at once (e.g., individual and group-level correspondence in the case of the "user-group" graph).

6.2 BIG-ALIGN: BIPARTITE GRAPH ALIGNMENT

Now that we have formulated the problem, we move on to the description of a technique to solve it. Our objective is two-fold: (i) In terms of effectiveness, given the non-convexity of Problem 6.3, our goal is to find a "good" local minimum; (ii) In terms of efficiency, we focus on carefully designing the search procedure. The two key idas of BiG-Align are:

- an alternating, projected gradient descent approach to find the local minima of the newly-defined optimization problem (Problem 6.3), and

- a series of optimizations: (a) a network-inspired initialization (Net-Init) of the correspondence matrices to find a good starting point, (b) automatic choice of the steps for the gradient descent, and (c) handling the node-multiplicity problem, i.e., the "problem" of having nodes with exactly the same structure (e.g., peripheral nodes of a star) to improve both effectiveness and efficiency.

Next, we start by building the core of our method, continue with the description of the three optimizations, and conclude with the pseudocode of the overall algorithm.

6.2.1 MATHEMATICAL FORMULATION

Following the standard approach in the literature, in order to solve the optimization problem (Problem 6.3), we propose to first relax the sparsity constraint, which is mathematically represented by the l_0-norm of the matrices' columns, and replace it with the l_1-norm,

$\sum_i |\mathbf{P}_i^{(v)}| = \sum_i \mathbf{P}_i^{(v)}$, where we also use the probabilistic constraint. Therefore, the sparsity constraint now takes the form: $\sum_{i,j} P_{ij} \leq t$ and $\sum_{i,j} Q_{ij} \leq t$. By using this relaxation and applying linear algebra operations, the bipartite graph alignment problem takes the following form.

Theorem 6.4 **[Augmented Cost Function]** *The optimization problem for the alignment of the bipartite graphs G_1 and G_2, with adjacency matrices $\mathbf{A}_1$ and $\mathbf{A}_2$, under the probabilistic and sparsity constraints (Problem 6.3), is equivalent to:*

$$\min_{\mathbf{P},\mathbf{Q}} f_{aug}(\mathbf{P},\mathbf{Q}) = \min_{\mathbf{P},\mathbf{Q}} \left\{ ||\mathbf{PA}_1\mathbf{Q} - \mathbf{A}_2||_F^2 + \lambda \sum_{i,j} P_{ij} + \mu \sum_{i,j} Q_{ij} \right\}$$
$$= \min_{\mathbf{P},\mathbf{Q}} \left\{ ||\mathbf{PA}_1\mathbf{Q}||_F^2 - 2\operatorname{Tr}\left(\mathbf{PA}_1\mathbf{QA}_2^T\right) + \lambda \mathbf{1}^T\mathbf{P}\mathbf{1} + \mu \mathbf{1}^T\mathbf{Q}\mathbf{1} \right\}, \qquad (6.1)$$

where $|| \bullet ||_F$ is the Frobenius norm of the enclosed matrix, $\mathbf{P}$ and $\mathbf{Q}$ are the user- and group-level correspondence matrices, and λ and μ are the sparsity penalties of $\mathbf{P}$ and $\mathbf{Q}$, respectively.

Proof. The minimization

$$\min_{\mathbf{P},\mathbf{Q}} ||\mathbf{PA}_1\mathbf{Q} - \mathbf{A}_2||_F^2 \qquad \text{(Problem 6.3)}$$

can be reduced to

$$\min_{\mathbf{P},\mathbf{Q}} \left\{ ||\mathbf{PA}_1\mathbf{Q}||_F^2 - 2\operatorname{Tr}\mathbf{PA}_1\mathbf{QA}_2^T \right\}.$$

Starting from the definition of the Frobenius norm of $\mathbf{PA}_1\mathbf{Q} - \mathbf{A}_2$ (Table 6.1), we obtain:

$$||\mathbf{PA}_1\mathbf{Q} - \mathbf{A}_2||_F^2 = \operatorname{Tr}\left(\mathbf{PA}_1\mathbf{Q} - \mathbf{A}_2\right)\left(\mathbf{PA}_1\mathbf{Q} - \mathbf{A}_2\right)^T$$
$$= \operatorname{Tr}\left(\mathbf{PA}_1\mathbf{Q}(\mathbf{PA}_1\mathbf{Q})^T - 2\mathbf{PA}_1\mathbf{QA}_2^T\right) + \operatorname{Tr}\left(\mathbf{A}_2\mathbf{A}_2^T\right)$$
$$= ||\mathbf{PA}_1\mathbf{Q}||_F^2 - 2\operatorname{Tr}\left(\mathbf{PA}_1\mathbf{QA}_2^T\right) + \operatorname{Tr}\left(\mathbf{A}_2\mathbf{A}_2^T\right),$$

where we used the property $\operatorname{Tr}\left(\mathbf{PA}_1\mathbf{QA}_2^T\right) = \operatorname{Tr}\left(\mathbf{PA}_1\mathbf{QA}_2^T\right)^T$. Given that the last term, $\operatorname{Tr}\left(\mathbf{A}_2\mathbf{A}_2^T\right)$, does not depend on $\mathbf{P}$ or $\mathbf{Q}$, it does not affect the minimization. $\square$

In summary, we solve the minimization problem by using a variant of the gradient descent algorithm. Given that the cost function in Equation (6.1) is bivariate, we use an alternating procedure to minimize it. We fix $\mathbf{Q}$ and minimize f_{aug} with respect to $\mathbf{P}$, and vice versa. If, during the two alternating minimization steps, the entries of the correspondence matrices become invalid temporarily, we use a projection technique to guarantee the probabilistic constraint: If $P_{ij} < 0$ or $Q_{ij} < 0$, we *project* the entry to 0. If $P_{ij} > 1$ or $Q_{ij} > 1$, we *project* it to 1. The update steps of the alternating, projected gradient descent approach (APGD) are given by the following theorem.

Theorem 6.5 **[Update Step]** *The update steps for the user- (P) and group-level (Q) correspondence matrices of APGD are given by:*

$$\mathbf{P}^{(k+1)} = \mathbf{P}^{(k)} - \eta_P \cdot \left(2(\mathbf{P}^{(k)}\mathbf{A}_1\mathbf{Q}^{(k)} - \mathbf{A}_2)\mathbf{Q}^{T^{(k)}}\mathbf{A}_1^T + \lambda\mathbf{11}^T\right)$$

$$\mathbf{Q}^{(k+1)} = \mathbf{Q}^{(k)} - \eta_Q \cdot \left(2\mathbf{A}_1^T\mathbf{P}^{T^{(k+1)}}(\mathbf{P}^{(k+1)}\mathbf{A}_1\mathbf{Q}^{(k)} - \mathbf{A}_2) + \mu\mathbf{11}^T\right),$$

where $\mathbf{P}^{(k)}$, $\mathbf{Q}^{(k)}$ are the correspondence matrices at iteration k, η_P and η_Q are the steps of the two phases of the APGD, and $\mathbf{1}$ is the all-1 column-vector.

Proof. The update steps for gradient descent are:

$$\mathbf{P}^{(k+1)} = \mathbf{P}^{(k)} - \eta_P \cdot \frac{\partial f_{aug}(\mathbf{P}, \mathbf{Q})}{\partial \mathbf{P}} \tag{6.2}$$

$$\mathbf{Q}^{(k+1)} = \mathbf{Q}^{(k)} - \eta_Q \cdot \frac{\partial f_{aug}(\mathbf{P}, \mathbf{Q})}{\partial \mathbf{Q}}, \tag{6.3}$$

where $f_{aug}(\mathbf{P}, \mathbf{Q}) = f(\mathbf{P}, \mathbf{Q}) + s(\mathbf{P}, \mathbf{Q})$, $f = ||\mathbf{P}\mathbf{A}_1\mathbf{Q} - \mathbf{A}_2||_F^2$, and $s(\mathbf{P}, \mathbf{Q}) = \lambda \sum_{i,j} P_{ij} + \mu \sum_{i,j} Q_{ij}$.

First, we compute the derivative of f with respect to $\mathbf{P}$ by using properties of matrix derivatives:

$$\frac{\partial f(\mathbf{P}, \mathbf{Q})}{\partial \mathbf{P}} = \frac{\partial\left(||\mathbf{P}\mathbf{A}_1\mathbf{Q}||_F^2 - 2\operatorname{Tr}\left(\mathbf{P}\mathbf{A}_1\mathbf{Q}\mathbf{A}_2^T\right)\right)}{\partial \mathbf{P}}$$

$$= \frac{\partial\operatorname{Tr}\left(\mathbf{P}\mathbf{A}_1\mathbf{Q}\mathbf{Q}^T\mathbf{A}_1^T\mathbf{P}^T\right)}{\partial \mathbf{P}} - 2\frac{\partial\operatorname{Tr}\left(\mathbf{P}\mathbf{A}_1\mathbf{Q}\mathbf{A}_2^T\right)}{\partial \mathbf{P}}$$

$$= 2(\mathbf{P}\mathbf{A}_1\mathbf{Q} - \mathbf{A}_2)\mathbf{Q}^T\mathbf{A}_1^T. \tag{6.4}$$

Then, by using properties of matrix derivatives and the invariant property of the trace under cyclic permutations $\operatorname{Tr}(\mathbf{P}\mathbf{A}_1\mathbf{Q}\mathbf{Q}^T\mathbf{A}_1^T\mathbf{P}) = \operatorname{Tr}(\mathbf{A}_1^T\mathbf{P}^T\mathbf{P}\mathbf{A}_1\mathbf{Q}\mathbf{Q}^T)$, we obtain the derivative of $f(\mathbf{P}, \mathbf{Q})$ with respect to $\mathbf{Q}$:

$$\frac{\partial f(\mathbf{P}, \mathbf{Q})}{\partial \mathbf{Q}} = \frac{\partial\left(||\mathbf{P}\mathbf{A}_1\mathbf{Q}||_F^2 - 2\operatorname{Tr}\mathbf{P}\mathbf{A}_1\mathbf{Q}\mathbf{A}_2^T\right)}{\partial \mathbf{Q}}$$

$$= \frac{\partial\operatorname{Tr}\left(\mathbf{P}\mathbf{A}_1\mathbf{Q}\mathbf{Q}^T\mathbf{A}_1^T\mathbf{P}^T\right) - 2\operatorname{Tr}\left(\mathbf{P}\mathbf{A}_1\mathbf{Q}\mathbf{A}_2^T\right)}{\partial \mathbf{Q}}$$

$$= \frac{\partial\operatorname{Tr}\left(\mathbf{A}_1^T\mathbf{P}^T\mathbf{P}\mathbf{A}_1\mathbf{Q}\mathbf{Q}^T\right)}{\partial \mathbf{Q}} - 2\frac{\partial\operatorname{Tr}\left(\mathbf{P}\mathbf{A}_1\mathbf{Q}\mathbf{A}_2^T\right)}{\partial \mathbf{Q}}$$

$$= \left(\mathbf{A}_1^T\mathbf{P}^T\mathbf{P}\mathbf{A}_1 + \left(\mathbf{A}_1^T\mathbf{P}^T\mathbf{P}\mathbf{A}_1\right)^T\right)\mathbf{Q} - 2\left(\mathbf{P}\mathbf{A}_1\right)^T\left(\mathbf{A}_2^T\right)^T$$

$$= 2\mathbf{A}_1^T\mathbf{P}^T\left(\mathbf{P}\mathbf{A}_1\mathbf{Q} - \mathbf{A}_2\right). \tag{6.5}$$

Finally, the partial derivatives of $s(\mathbf{P}, \mathbf{Q})$ with respect to $\mathbf{P}$ and $\mathbf{Q}$ are

$$\frac{\partial s(\mathbf{P}, \mathbf{Q})}{\partial \mathbf{P}} = \frac{\partial\left(\mathbf{1}^T\mathbf{P}\mathbf{1} + \mathbf{1}^T\mathbf{Q}\mathbf{1}\right)}{\partial \mathbf{P}} = \mathbf{11}^T. \tag{6.6}$$

$$\frac{\partial s(\mathbf{P}, \mathbf{Q})}{\partial \mathbf{Q}} = \frac{\partial \left(\mathbf{1}^T \mathbf{P} \mathbf{1} + \mathbf{1}^T \mathbf{Q} \mathbf{1}\right)}{\partial \mathbf{P}} = \mathbf{1} \mathbf{1}^T. \tag{6.7}$$

By substituting Equations (6.4) and (6.6) in Equation (6.2), we obtain the update step for $\mathbf{P}$. Similarly, by substituting Equations (6.5) and (6.7) in Equation (6.3), we get the update step for $\mathbf{Q}$.   □

We note that the assumption in the above formulas is that $\mathbf{A}_1$ and $\mathbf{A}_2$ are rectangular, adjacency matrices of bipartite graphs. It turns out that this formulation has a nice connection to the standard formulation for unipartite graph matching if we treat the input bipartite graphs as unipartite (i.e., symmetric, square, adjacency matrix). We summarize this equivalence in the following proposition.

Proposition 6.6   [Equivalence to Unipartite Graph Alignment] If the rectangular adjacency matrices of the bipartite graphs are converted to square matrices, then the minimization is done with respect to the coupled matrix $\mathbf{P}^*$:

$$\mathbf{P}^* = \begin{pmatrix} \mathbf{P} & \mathbf{0} \\ \mathbf{0} & \mathbf{Q} \end{pmatrix}.$$

That is, Problem 6.3 becomes $\min_{\mathbf{P}^*} ||\mathbf{P}^* \mathbf{A}_1 \mathbf{P}^{*T} - \mathbf{A}_2||_F^2$, which is equivalent to the unipartite graph problem introduced at the beginning of Section 6.1.

6.2.2   PROBLEM-SPECIFIC OPTIMIZATIONS

Up to this point, we have the mathematical foundation at our disposal to build our algorithm, BIG-ALIGN. But first we have to make three design decisions.
(D1) How to initialize the correspondence matrices?
(D2) How to choose the steps for the APGD?
(D3) How to handle structurally equivalent nodes?
    The baseline approach, which we will refer to as BIG-ALIGN-BASIC, consists of the simplest answers to these questions: (D1) uniform initialization of the correspondence matrices, (D2) "small," constant step for the gradient descent, and (D3) no specific manipulation of the structurally equivalent nodes. Next, we elaborate on sophisticated choices for the initialization and optimization step that render our algorithm more efficient. We also introduce the "node-multiplicity" problem, i.e., the problem of structurally equivalent nodes, and propose a way to deal with it.

(D1) How to initialize the correspondence matrices?
    The optimization problem is non-convex (not even bi-convex), and the gradient descent gets stuck in local minima, depending heavily on the initialization. There are several different

ways of initializing the correspondence matrices $\mathbf{P}$ and $\mathbf{Q}$, such as random, degree-based, and eigenvalue-based [56, 212]. While each of these initializations has its own rationality, they are designed for unipartite graphs and hence ignore the skewness of the real, large-scale bipartite graphs. To address this issue, we propose a network-inspired approach (NET-INIT), which is based on the following observation about large-scale, real biparite graphs:

Observation 6.7   Large, real networks have skewed or power-law-like degree distribution [9, 36, 64]. Specifically in bipartite graphs, usually one of the node sets is significantly smaller than the other, and has skewed degree distribution.

The implicit assumption[2] of NET-INIT is that a person is almost equally popular in different social networks, or, more generally, an entity has similar "behavior" across the input graphs. In our work, we have found that such behavior can be well captured by the node degree. However, the technique we describe below can be naturally applied to other features (e.g., weight, ranking, clustering coefficient) that may capture the node behavior better.

Our initialization approach consists of four steps. For the description of our approach, we refer to the example of the LinkedIn and Facebook bipartite graphs, where the first set consists of users, and the second set of groups. For the sake of the example, we assume that the set of groups is significantly smaller than the set of users.

Step 1   **Match 1-by-1 the top-k** high-degree groups in the LinkedIn and Facebook graphs. To find k, we borrow the idea of the scree plot, which is used in Principal Component Analysis (PCA): we sort the unique degrees of each graph in descending order, and create the plot of unique degree vs. rank of node (Figure 6.1a). In this plot, we detect the "knee" and up to the corresponding degree we "safely" match the groups of the two graphs one-by-one, i.e., the most popular group of LinkedIn is aligned initially with the most popular group of Facebook, etc. For the automatic detection of the knee, we consider the plot piecewise, and assume that the knee occurs when the slope of a line segment is less than 5% of the slope of the previous segment.

Step 2   For each of the matched groups, we propose to **align their neighbors** based on their Relative Degree Difference (RDD).

Definition 6.8 RDD.   The Relative Degree Distance (RDD) function that aligns node i of graph $\mathbf{A}_1$ to node j of $\mathbf{A}_2$ is:

$$\text{rdd}(i, j) = \left(1 + \frac{|\deg(i) - \deg(j)|}{(\deg(i) + \deg(j))/2}\right)^{-1}, \tag{6.8}$$

where $\deg(\bullet)$ is the degree of the corresponding node.

---

[2]If the assumption does not hold, no method is guaranteed to find the alignment based purely on the structure of the graphs, but they can still reveal similarities between nodes.

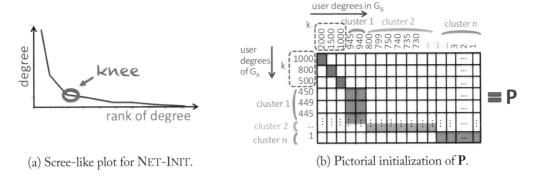

(a) Scree-like plot for NET-INIT.   (b) Pictorial initialization of **P**.

Figure 6.1: (a) Choice of k in Step 1 of NET-INIT. (b) Initialization of the node/user-level correspondence matrix by NET-INIT.

The idea behind this approach is that a node in one graph more probably corresponds to a node with similar degree in another graph, than to a node with very different degree. The above function assigns higher probabilities to alignments of similar nodes, and lower probabilities to alignments of very dissimilar nodes with respect to their degrees.

We note that the RDD function, rdd(i, j), corresponds to the degree-based similarity between node i and node j. However, it can be generalized to other properties that the nodes are expected to share across different graphs. Equation (6.8) captures one additional desired property: it penalizes the alignments based on the relative difference of the degrees. For example, two nodes of degrees 1 and 20, respectively, are less similar than two nodes with degrees 1,001 and 1,020.

Step 3   Create c_g **clusters of the remaining groups** in both networks, based on their degrees. **Align the clusters 1-by-1** according to the degrees (e.g., "high," "low"), and initialize the correspondences *within* the matched clusters using the RDD.

Step 4   Create c_u **clusters of the remaining users** in both networks, based on their degrees. Align the users using the RDD approach within the corresponding user clusters.
(D2) How to choose the steps for the APGD method?
One of the most important parameters of the APGD method is η (the step of approaching the minimum point), which determines its convergence rate. In an attempt to automatically determine the step, we use the *line search* approach [33], which is described in Algorithm 6.2. Line search is a strategy that finds the local optimum for the step. Specifically, in the first phase of APGD, line search determines η_P by treating the objective function, f_{aug}, as a function of η_P (instead of a function of **P** or **Q**) and loosely minimizing it. In the second phase of APGD, η_Q is determined similarly. Next we introduce three variants of our method that differ in the way the steps are computed.

Variant 1: BiG-Align-Points. Our first approach consists of approximately minimizing the augmented cost function: we randomly pick some values for η_P within some "reasonable" range, and compute the value of the cost function. We choose the step η_P that corresponds to the minimum value of the cost function. We define η_Q similarly. This approach is computationally expensive, as we shall see in Section 6.4.

Variant 2: BiG-Align-Exact. By carefully handling the objective function of our optimization problem, we can find the closed (exact) forms for η_P and η_Q, which are given in the next theorem.

Theorem 6.9 **[Optimal Step Size for P]** *In the first phase of APGD, the value of the step η_P that exactly minimizes the augmented function, $f_{aug}(\eta_P)$, is given by:*

$$\eta_P = \frac{2\,\mathrm{Tr}\left\{(\mathbf{P}^{(k)}\mathbf{A}_1\mathbf{Q})(\Delta_\mathbf{P}\mathbf{A}_1\mathbf{Q})^T - (\Delta_\mathbf{P}\mathbf{A}_1\mathbf{Q})\mathbf{A}_2^T\right\} + \lambda \sum_{i,j} \Delta_{Pij}}{2||\Delta_P\mathbf{A}_1\mathbf{Q}||_F^2}, \tag{6.9}$$

where $\mathbf{P}^{(k+1)} = \mathbf{P}^{(k)} - \eta_P \Delta_\mathbf{P}$, $\Delta_\mathbf{P} = \nabla_\mathbf{P} f_{aug}|_{\mathbf{P}=\mathbf{P}^{(k)}}$ and $\mathbf{Q} = \mathbf{Q}^{(k)}$.

Proof. To find the step η_P that minimizes $f_{aug}(\eta_P)$, we take its derivative and set it to 0:

$$\frac{df_{aug}}{d\eta_P} = \frac{d\left(\mathrm{Tr}\left\{\mathbf{P}^{(k+1)}\mathbf{A}_1\mathbf{Q}(\mathbf{P}^{(k+1)}\mathbf{A}_1\mathbf{Q})^T - 2\mathbf{P}^{(k+1)}\mathbf{A}_1\mathbf{Q}\mathbf{A}_2^T\right\} + \lambda \sum_{i,j} P_{ij}^{(k+1)}\right)}{d\eta_P} = 0, \tag{6.10}$$

where $\mathbf{P}^{(k+1)} = \mathbf{P}^{(k)} - \eta_P \Delta_P$, where $\Delta_P = \nabla_\mathbf{P} f_{aug}|_{\mathbf{P}=\mathbf{P}^{(k)}}$. It also holds that

$$\mathrm{Tr}\,(\mathbf{P}^{(k+1)}\mathbf{A}_1\mathbf{Q}(\mathbf{P}^{(k+1)}\mathbf{A}_1\mathbf{Q})^T) - 2\mathbf{P}^{(k+1)}\mathbf{A}_1\mathbf{Q}\mathbf{A}_2^T) =$$
$$||\mathbf{P}^{(k)}\mathbf{A}_1\mathbf{Q}||_F^2 - 2\,\mathrm{Tr}\,\mathbf{P}^{(k)}\mathbf{A}_1\mathbf{Q}\mathbf{A}_2^T + \eta_P^2||\Delta_P\mathbf{A}_1\mathbf{Q}||_F^2 +$$
$$+2\eta_P\,\mathrm{Tr}\,(\Delta_P\mathbf{A}_1\mathbf{Q}\mathbf{A}_2^T) - 2\eta_P\,\mathrm{Tr}\,(\mathbf{P}^{(k)}\mathbf{A}_1\mathbf{Q})(\Delta_P\mathbf{A}_1\mathbf{Q}) \tag{6.11}$$

Substituting Equation (6.11) in Equation (6.10), and solving for η_P yields the "best value" of η_P as defined by the line search method. □

Similarly, we find the appropriate value for the step η_Q of the second phase of APGD.

Theorem 6.10 **[Optimal Step Size for Q]** *In the second phase of APGD, the value of the step η_Q that exactly minimizes the augmented function, $f_{aug}(\eta_Q)$, is given by:*

$$\eta_Q = \frac{2\,\mathrm{Tr}\left\{(\mathbf{PA}_1\mathbf{Q}^{(k)})(\mathbf{PA}_1\Delta_Q)^T - (\mathbf{PA}_1\Delta_Q)\mathbf{A}_2^T\right\} + \mu \sum_{i,j} \Delta_{Qij}}{2||\mathbf{PA}_1\Delta_Q||_F^2}, \tag{6.12}$$

where $\Delta_\mathbf{Q} = \nabla_\mathbf{Q} f_{aug}|_{\mathbf{Q}=\mathbf{Q}^{(k)}}$, $\mathbf{P} = \mathbf{P}^{(k)}$, and $\mathbf{Q}^{(k+1)} = \mathbf{Q}^{(k)} - \eta_Q \Delta_\mathbf{Q}$.

Proof. The computations for η_Q is symmetric to the computation of η_P (Theorem 6.9), and thus we omit it. $\qquad\square$

BiG-Align-Exact is significantly faster than BiG-Align-Points. It turns out that we can increase the efficiency even more, as experimentation with real data revealed that the values of the gradient descent steps that minimize the objective function do not change drastically in every iteration (Figure 6.2). This led to the third variation of our algorithm.

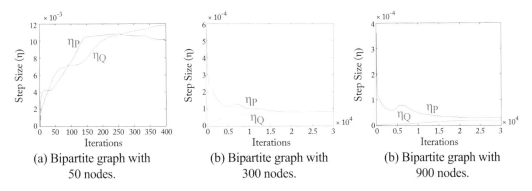

Figure 6.2: Hint for speedup: Size of optimal step for **P** (blue) and **Q** (green) vs. the number of iterations. We observe that the optimal step sizes do not change dramatically in consecutive iterations, and, thus, skipping some computations almost does not affect the accuracy at all.

Variant 3: BiG-Align-Skip. This variation applies exact line search for the first few (e.g., 100) iterations, and then updates the values of the steps every few (e.g., 500) iterations. This significantly reduces the computations for determining the optimal step sizes.

(D3) How to handle structurally equivalent nodes?

One last observation that renders BiG-Align more efficient is the following.

Observation 6.11 In the majority of graphs, there is a significant number of nodes that cannot be distinguished, because they have exactly the same structural features.

For instance, in many real-world networks, a commonplace structure is stars [103], but it is impossible to tell the peripheral nodes apart. Other examples of non-distinguishable nodes include the members of cliques, and full bipartite cores (Chapter 2, [124]).

To address this problem, we introduce a pre-processing phase at which we eliminate nodes with identical structures by aggregating them in super-nodes. For example, a star with 100 peripheral nodes which are connected to the center by edges of weight 1, will be replaced by a super-node connected to the central node of the star by an edge of weight 100. This subtle step not only leads to a better optimization solution, but also improves the efficiency by reducing the scale of graphs that are actually fed into our BiG-Align.

6.2.3   ALGORITHM DESCRIPTION

The previous subsections shape up the proposed algorithm, BiG-Align, the pseudocode of which is given in Algorithms 6.1 and 6.2.

In our implementation, the only parameter that the user is required to input is the sparsity penalty, λ. The bigger this parameter is, the more entries of the matrices are forced to be 0. We set the other sparsity penalty $\mu = \frac{\lambda * (\text{elements in } \mathbf{Q})}{\text{elements in } \mathbf{P}}$, so that the penalty per non-zero element of $\mathbf{P}$ and $\mathbf{Q}$ is the same.

It is worth mentioning that, in contrast to the approaches found in the literature, our method does not use the classic Hungarian algorithm to find the hard correspondences between the nodes of the bipartite graphs. Instead, we rely on a fast approximation: we align each row i (node/user) of $\mathbf{P}^T$ with the column j (node/user) that has the maximum probability. It is clear that this assignment is very fast, and even parallelizable, as each node alignment can be processed independently. Moreover, it allows aligning multiple nodes of one graph with the same node of the other graph, a property that is desirable especially in the case of structurally equivalent nodes.

Figure 6.3 depicts how the cost and accuracy of the alignment change with respect to the number of iterations of the gradient descent algorithm.

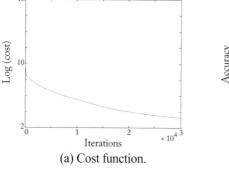
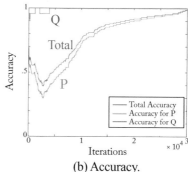

(a) Cost function.

(b) Accuracy.

Figure 6.3: BiG-Align (900 nodes, $\lambda = 0.1$): As desired, the cost of the objective function drops with the number of iterations, while the accuracy both on node- (green) and community-level (red) increases. The exact definition of accuracy is given in Section 6.4.1.

6.3   UNI-ALIGN: EXTENSION TO UNIPARTITE GRAPH ALIGNMENT

Although our primary target for BiG-Align is bipartite graphs (which by themselves already stand for a significant portion of real graphs), as a side-product, BiG-Align also offers an alternative, fast solution to the alignment problem of unipartite graphs. Our approach consists of two steps.

Algorithm 6.1 BiG-Align-Exact: Bipartite Graph Alignment

Input : $\mathbf{A}_1$, $\mathbf{A}_2$, λ, MAXITER, $\epsilon = 10^{-6}$; $cost(0) = 0$; $k = 1$;
Output : The correspondence matrices $\mathbf{P}$ and $\mathbf{Q}$

1: /* STEP 1: pre-processing for node-multiplicity */
2: aggregating identical nodes
3: /* STEP 2: initialization */
4: [P0, Q0] = Net-Init
5: $cost(1) = f_{aug}(\mathbf{P0}, \mathbf{Q0})$
6: /* STEP 3: alternating projected gradient descent (APGD) */
7: **while** $|cost(k-1) - cost(k)|/cost(k-1) > \epsilon$ AND $k <$ MAXITER **do**
8:   $k + +$
9:   /* Phase 1: fixed $\mathbf{Q}$, minimization with respect to $\mathbf{P}$ */
10:   $\eta_{Pk} = $ LineSearch_P$(\mathbf{P}^{(k)}, \mathbf{Q}^{(k)}, \nabla_{\mathbf{P}} f_{aug}|_{\mathbf{P}=\mathbf{P}^{(k)}})$
11:   $\mathbf{P}^{(k+1)} = \mathbf{P}^{(k)} - \eta_{Pk} \nabla_{\mathbf{P}} f_{aug}(\mathbf{P}^{(k)}, \mathbf{Q}^{(k)})$
12:   validProjection$(\mathbf{P}^{(k+1)})$
13:   /* Phase 2: fixed $\mathbf{P}$, minimization with respect to $\mathbf{Q}$ */
14:   $\eta_{Qk} = $ LineSearch_Q$(\mathbf{P}^{(k+1)}, \mathbf{Q}^{(k)}, \nabla_{\mathbf{Q}} f_{aug}|_{\mathbf{Q}=\mathbf{Q}^{(k)}})$
15:   $\mathbf{Q}^{(k+1)} = \mathbf{Q}^{(k)} - \eta_{Qk} \nabla_{\mathbf{Q}} f_{aug}(\mathbf{P}^{(k+1)}, \mathbf{Q}^{(k)})$
16:   validProjection$(\mathbf{Q}^{(k+1)})$
17:   $cost(k) = f_{aug}(\mathbf{P}, \mathbf{Q})$
18: **end while**
19:
20: **return** $\mathbf{P}^{(k+1)}$, $\mathbf{Q}^{(k+1)}$

21: /* Projection Step */
22: **function** validProjection P
23: **for** $\mathbf{P}_{ij}$ in $\mathbf{P}$ **do**
24:   **if** $\mathbf{P}_{ij} < 0$ **then**
25:     $\mathbf{P}_{ij} = 0$
26:   **else if** $\mathbf{P}_{ij} > 1$ **then**
27:     $\mathbf{P}_{ij} = 1$
28:   **end if**
29: **end for**
30: **end function**

Step 1: Uni- to Bipartite Graph Conversion   The first step involves converting the $n \times n$ unipartite graphs to bipartite graphs. Specifically, we can first extract d node features, such as degree, edges in a node's egonet (= induced subgraph of the node and its neighbors), and clustering coefficient. Then, we can form the $n \times d$ bipartite graph node-to-feature, where $n \gg d$. The runtime of this step depends on the time complexity of extracting the selected features.

Step 2: Finding P   We note that in this case, the alignment of the feature sets of the bipartite graphs is known, i.e., $\mathbf{Q}$ is an identity matrix, since we extract the same type of features from

---

Algorithm 6.2 Line search for η_P and η_Q

Input : $\mathbf{P}, \mathbf{Q}, \Delta_{\mathbf{P}}, \Delta_{\mathbf{Q}}$
Output : η_P, η_Q

1: **function** LineSearch_P ($\mathbf{P}, \mathbf{Q}, \Delta_{\mathbf{P}}$)
2: **return**

$$\eta_P = \frac{2\,\mathrm{Tr}\left\{(\mathbf{P}^{(k)}\mathbf{A}_1\mathbf{Q})(\Delta_{\mathbf{P}}\mathbf{A}_1\mathbf{Q})^T - (\Delta_{\mathbf{P}}\mathbf{A}_1\mathbf{Q})\mathbf{A}_2^T\right\} + \lambda\sum_{i,j}\Delta_{Pij}}{2\|\Delta_P\mathbf{A}_1\mathbf{Q}\|_F^2}$$

3: **end function**

4: **function** LineSearch_Q ($\mathbf{P}, \mathbf{Q}, \Delta_{\mathbf{Q}}$)
5: **return**

$$\eta_Q = \frac{2\,\mathrm{Tr}\left\{(\mathbf{P}\mathbf{A}_1\mathbf{Q}^{(k)})(\mathbf{P}\mathbf{A}_1\Delta_{Q})^T - (\mathbf{P}\mathbf{A}_1\Delta_{\mathbf{Q}})\mathbf{A}_2^T\right\} + \mu\sum_{i,j}\Delta_{Qij}}{2\|\mathbf{P}\mathbf{A}_1\Delta_{Q}\|_F^2}$$

6: **end function**

---

the graphs. Thus, we only need to align the n nodes, i.e., compute $\mathbf{P}$. We revisit Equation (6.1) of our initial minimization problem, and now we want to minimize it only with respect to $\mathbf{P}$. By setting the derivative of f_{aug} with respect to $\mathbf{P}$ equal to 0, we have:

$$\mathbf{P}\cdot\left(\mathbf{A}_1\mathbf{A}_1^T\right) = \mathbf{A}_2\mathbf{A}_1^T - \lambda/2\cdot\mathbf{1}\mathbf{1}^T,$$

where $\mathbf{A}_1$ is a $n \times d$ matrix. If we do SVD (Singular Value Decomposition) on this matrix, i.e., $\mathbf{A}_1 = \mathbf{USV}$, the Moore-Penrose pseudo-inverse of $\mathbf{A}_1\mathbf{A}_1^T$ is $(\mathbf{A}_1\mathbf{A}_1^T)^\dagger = \mathbf{US}^{-2}\mathbf{U}^T$. Therefore, we have

$$\begin{aligned}
\mathbf{P} &= \left(\mathbf{A}_2\mathbf{A}_1^T - \lambda/2\mathbf{1}\mathbf{1}^T\right)\left(\mathbf{A}_1\mathbf{A}_1^T\right)^\dagger \\
&= \left(\mathbf{A}_2\mathbf{A}_1^T - \lambda/2\mathbf{1}\mathbf{1}^T\right)\left(\mathbf{US}^{-2}\mathbf{U}^T\right) \\
&= \mathbf{A}_2\cdot\left(\mathbf{A}_1^T\mathbf{US}^{-2}\mathbf{U}^T\right) - \mathbf{1}\cdot\left(\lambda/2\cdot\mathbf{1}^T\mathbf{US}^{-2}\mathbf{U}^T\right) \\
&= \mathbf{A}_2\cdot\mathbf{X} - \mathbf{1}\cdot\mathbf{Y}
\end{aligned} \tag{6.13}$$

where $\mathbf{X} = \mathbf{A}_1^T\mathbf{US}^{-2}\mathbf{U}^T$ and $\mathbf{Y} = \lambda/2\cdot\mathbf{1}^T\mathbf{US}^{-2}\mathbf{U}^T$. Hence, we can exactly (*non*-iteratively) find $\mathbf{P}$ from Equation (6.13). It can be shown that the time complexity for finding $\mathbf{P}$ is $\mathbf{O}(nd^2)$ (after omitting the simpler terms), which is linear on the number of nodes of the input graphs.

What is more, we can see from Equation (6.13) that $\mathbf{P}$ itself has the low-rank structure. In other words, we do not need to store $\mathbf{P}$ in the form of $n \times n$. Instead, we can represent (compress) $\mathbf{P}$ as the multiplication of two low-rank matrices $\mathbf{X}$ and $\mathbf{Y}$, whose additional space cost is just $O(nd + n) = O(nd)$.

6.4   EMPIRICAL RESULTS

In this section, we perform experiments to evaluate the methods BiG-Align and Uni-Align and answer the following questions.

Q1. How does BiG-Align and its variants fare in terms of alignment accuracy and runtime, and how do they compare to state-of-the-art approaches?

Q2. How does its extension, Uni-Align, perform on unipartite graphs in terms of alignment accuracy and runtime?

Q3. How do the methods scale with the number of edges in the input graphs?

The code for all the methods is written in Matlab and the experiments were run on Intel(R) Xeon(R) CPU 5160 @ 3.00 GHz, with 16 GB RAM memory.

Baseline Methods   To the best of our knowledge, no graph matching algorithm has been designed for bipartite graphs. Throughout this section, we compare our algorithms to 3 state-of-the-art approaches, which are succinctly described in Table 6.2: (i) Umeyama, the influential eigenvalue decomposition-based approach proposed by Umeyama [212]; (ii) NMF-based, a recent approach based on Non-negative Matrix Factorization [56]; and (iii) NetAlign-full and NetAlign-deg, two variations of a fast, and scalable Belief Propagation-based (BP) approach [23]. Some details about these approaches are provided in the related work (Section 6.6).

In order to apply these approaches to bipartite graphs, we convert the latter to unipartite by using Lemma 6.6. In addition to that, since the BP-based approach requires not only the two input graphs, but also a bipartite graph that encodes the possible matchings per node, we use two heuristics to form the required bipartite "matching" graph: (a) full bipartite graph, which essentially conveys that we have no domain information about the possible alignments, and each node of the first graph can be aligned with *any* node of the second graph (NetAlign-full); and (b) degree-based bipartite graph, where only nodes with the same degree in both graphs are considered possible matchings (NetAlign-deg).

6.4.1   ACCURACY AND RUNTIME OF BIG-ALIGN

For the experiments on bipartite graphs, we use the movie-genre graph of the MovieLens network.[3] Each of the 1,027 movies is linked to at least one of the 23 genres (e.g., comedy, romance, drama). Specifically, from this network, we extract subgraphs of different sizes. Then, following the tradition in the literature [56], for each of the subgraphs we generate permutations, $\mathbf{A}_2$, with *noise* from 0%–20% using the formula $\mathbf{A}_{2ij} = (\mathbf{P}\mathbf{A}_1\mathbf{Q})_{ij} \cdot (1 + noise * r_{ij})$, where r_{ij} is a random number in [0, 1]. For each noise level and graph size, we generate 10 distinct permutations of the initial subnetwork. We run the alignment algorithms on all the pairs of the original and per-

³http://www.movielens.org

Table 6.2: Graph alignment algorithms: name conventions, short description, type of graphs for which they were designed ("uni-" for unipartite, "bi-" for bipartite graphs), and reference

Name	Description	Graph	Source
Umeyama	Eigenvalue-based	uni-	[221]
NMF-based	NMF-based	uni-	[59]
NetAlign-full	BP-based with uniform init.	uni-	Modified
NetAlign-deg	BP-based with same-degree init.	uni-	From [24]
BIG-ALIGN-Basic	APGD (no optimizations)	bi-	Current
BIG-ALIGN-Points	APGD + approx. Line Search	bi-	Current
BIG-ALIGN-Exact	APGD + exact Line Search	bi-	Current
BIG-ALIGN-Skip	APGD + skip some Line Search	bi-	Current
UNI-ALIGN	BIG-ALIGN-inspired (SVD)	uni-	Current

muted subgraphs, and report the mean accuracy and runtime. For all the variants of BIG-ALIGN, we set the sparsity penalty $\lambda = 0.1$.

How do we compute the accuracy of the methods? For the state-of-the-art methods, which find "hard" alignments between the nodes, the accuracy is computed as usual: only if the true correspondence is found, the corresponding matching is deemed correct. In other words, we use the state-of-the-art algorithms off-the-shelf. For our method, BIG-ALIGN, which has the advantage of finding "soft," probabilistic alignments, we consider two cases for evaluating its accuracy: (i) *Correct Alignment.* If the true correspondence coincides with the most probable matching, we count the node alignment as correct; and (ii) *Partially Correct Alignment.* If the true correspondence is *among* the most probable matchings (tie), the alignment thereof is deemed partially correct and weighted by (# of nodes in tie)/ (total # of nodes).

Accuracy   Figures 6.4a and 6.4b present the accuracy of the methods for two different graph sizes and *varying level of noise* in the permutations. We observe that BIG-ALIGN outperforms all the other methods in most cases with a large margin. In Figure 6.4b, the only exception is the case of 20% of noise in the 900-nodes graphs where NetAlign-deg and NetAlign-full perform slightly better than our algorithm, BIG-ALIGN-Exact. The results for other graph sizes are along the same lines, and therefore are omitted for space.

Figure 6.5a depicts the accuracy of the alignment methods *for varying graph size*. For graphs with different sizes, the variants of our method achieve significantly higher accuracy (70–98%) than the baselines (10–58%). Moreover, surprisingly, BIG-ALIGN-Skip performs slightly better than BIG-ALIGN-Exact, although the former skips several updates of the gradient descent steps. The only exception is for the smallest graph size, where the consecutive optimal steps change significantly (Figure 6.2a), and, thus, skipping computations affects the performance.

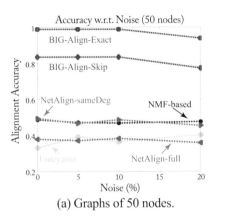

(a) Graphs of 50 nodes.

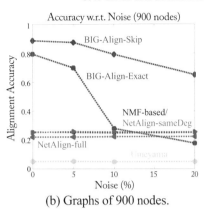

(b) Graphs of 900 nodes.

Figure 6.4: [Higher is better.] Accuracy of bipartite graph alignment vs. level of noise (0–20%). BiG-Align-Exact (red line with square marker), almost always, outperforms the baseline methods.

NetAlign-full and Umeyama's algorithm are the least accurate methods, while NMF-based and NetAlign-deg achieve medium accuracy. Finally, the accuracy vs. runtime plot in Figure 6.5c shows that our algorithms have two desired properties: they achieve *better* performance, *faster* than the baseline approaches.

Runtime   Figure 6.5b presents the runtime as a function of the number of edges in the graphs. Umeyama's algorithm and NetAlign-deg are the fastest methods, but at the cost of accuracy; BiG-Align is up to 10× more accurate in the cases that it performs slower. The third best method is BiG-Align-Skip, closely followed by BiG-Align-Exact. BiG-Align-Skip is up to 174× faster than the NMF-based approach, and up to 19× faster than NetAlign-full. However, our simplest method that uses line search, BiG-Align-Points, is the slowest approach and given that it takes too long to terminate for graphs with more than 1.5 K edges, we omit several data points in the plot.

We note that BiG-Align is a single machine implementation, but it has the potential for further speed-up. For example, it could be parallelized by splitting the optimization problem to smaller subproblems (by decomposing the matrices, and doing simple column-row multiplications). Moreover, instead of the basic gradient descent algorithm, we can use a variant method, the stochastic gradient descent, which is based on sampling.

Variants of BiG-Align   Before we continue with the evaluation of Uni-Align, we present in Table 6.3 the runtime and accuracy of all the variants of BiG-Align for aligning movie-genre graphs with varying sizes and permutations with noise level 10%. The parameters used in this experiment are $\epsilon = 10^{-5}$, and $\lambda = 0.1$. For BiG-Align-Basic, η is constant and equal to 10^{-4}, while the correspondence matrices are initialized uniformly. This is not the best setting for all

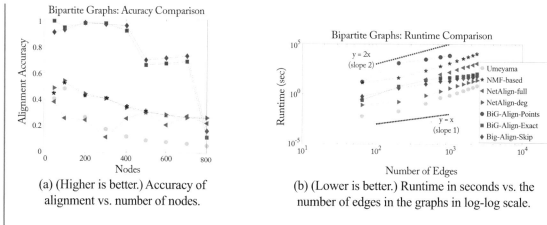

(a) (Higher is better.) Accuracy of alignment vs. number of nodes.

(b) (Lower is better.) Runtime in seconds vs. the number of edges in the graphs in log-log scale.

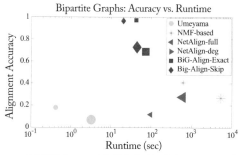

(c) (Higher and left is better.) Accuracy of alignment vs. runtime in seconds for graphs with 300 nodes (small markers) and 700 nodes (big markers).

Figure 6.5: Accuracy and runtime of alignment of bipartite graphs. (a) BiG-Align-Exact and BiG-Align-Skip (red lines) significantly outperform all the alignment methods for almost all the graph sizes, in terms of accuracy; (c) BiG-Align-Exact and BiG-Align-Skip (red squares/ovals) are faster and more accurate than the baselines for both graph sizes. (b) The BiG-Align variants are faster than all the baseline approaches, except for Umeyama's algorithm.

the pairs of graphs that we are aligning, and it results in very low accuracy. On the other hand, BiG-Align-Skip is not only $\sim 350\times$ faster than BiG-Align-Points, but also more accurate. Moreover, it is $\sim 2\times$ faster than BiG-Align-Exact with higher or equal accuracy. The speedup can be further increased by skipping more updates of the gradient descent steps.

Overall, the results show that a naive solution of the optimization problem, such as BiG-Align-Basic, is not sufficient, and the optimizations we propose in Section 6.2 are crucial and render our algorithm efficient.

Table 6.3: Runtime (top) and accuracy (bottom) comparison of the BiG-Align variants: BiG-Align-Basic, BiG-Align-Points, BiG-Align-Exact, and BiG-Align-Skip. BiG-Align-Skip is not only faster, but also comparably or more accurate than BiG-Align-Exact.

Nodes	BiG-Align-Basic		BiG-Align-Points		BiG-Align-Exact		BiG-Align-Skip	
	Mean	Std	Mean	Std	Mean	Std	Mean	Std
Runtime (sec)								
50	0.07	0.00	17.3	0.05	0.24	0.08	0.56	0.01
100	0.023	0.00	1245.7	394.55	5.6	2.93	3.9	0.05
200	31.01	16.58	2982.1	224.81	25.5	0.39	10.1	0.10
300	0.032	0.00	5240.9	30.89	42.1	1.61	20.1	1.62
400	0.027	0.01	7034.5	167.08	45.8	2.058	21.3	0.83
500	0.023	0.01	-	-	57.2	2.22	36.6	0.60
600	0.028	0.01	-	-	64.5	2.67	40.8	1.26
700	0.029	0.01	-	-	73.6	2.78	44.6	1.23
800	166.7	1.94	-	-	86.9	3.63	49.9	1.06
900	211.9	5.30	-	-	111.9	2.96	61.8	1.28
Accuracy								
50	0.071	0.00	0.982	0.02	0.988	0	0.904	0.03
100	0.034	0.00	0.922	0.07	0.939	0.06	0.922	0.07
200	0.722	0.37	0.794	0.01	0.973	0.01	0.975	0.00
300	0.014	0.00	0.839	0.02	0.972	0.01	0.964	0.01
400	0.011	0.00	0.662	0.02	0.916	0.03	0.954	0.01
500	0.011	0.00	-	-	0.66	0.20	0.697	0.24
600	0.005	0.00	-	-	0.67	0.20	0.713	0.23
700	0.004	0.00	-	-	0.69	0.20	0.728	0.19
800	0.013	0.00	-	-	0.12	0.02	0.165	0.03
900	0.015	0.00	-	-	0.17	0.20	0.195	0.22

6.4.2   ACCURACY AND RUNTIME OF UNI-ALIGN

To evaluate our proposed method, Uni-Align, for aligning unipartite graphs, we use the Facebook who-links-to-whom graph [215], which consists of approximately 64 K nodes. In this case, the baseline approaches are readily employed, while our method requires the conversion of the given unipartite graph to bipartite. We do so by extracting some unweighted egonet[4] features for each node (degree of node, degree of egonet,[5] edges of egonet, mean degree of the node's neighbors). As before, from the initial graph we extract subgraphs of size 100–800 nodes (or equivalently, 264–266 K edges), and create 10 noisy permutations (per noise level) as before.

Accuracy   The accuracy vs. runtime plot in Figure 6.6a shows that UNI-ALIGN outperforms all the other methods in terms of accuracy and runtime for all the graph sizes depicted. Although NMF achieves a reasonably good accuracy for the graph of 200 nodes, it takes too long to terminate; we stopped the runs for graphs of bigger sizes as the execution was taking too long. The remaining approaches are fast enough, but yield poor accuracy.

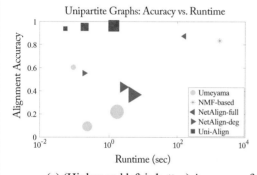

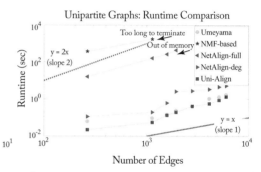

(a) (Higher and left is better.) Accuracy of alignment vs. runtime in seconds for Facebook friendship subgraphs of size 200 (small markers), 400 (medium markers), and 800 (big markers).

(b) (Lower is better.) Runtime in seconds vs. the number of edges in log-log scale.

Figure 6.6: Accuracy and runtime of alignment of unipartite graphs. (a) UNI-ALIGN (red points) is more accurate and faster than all the baselines for all graph sizes. `NetAlign-full` and `NMF-based` either run out of memory or take too long to terminate for medium sized graphs with more than 1.5–2.8 K edges. (b) UNI-ALIGN (red squares) is faster than all the baseline approaches, followed closely by `Umeyama`'s approach (green circles).

Runtime   Figure 6.6b compares the graph alignment algorithms with respect to their running time (in logscale). UNI-ALIGN is the fastest approach, closely followed by Umeyama's algorithm. `NetAlign-deg` is some orders of magnitude slower than the previously mentioned methods. However, `NetAlign-full` ran out of memory for graphs with more than 2.8 K edges; we stopped the runs of the `NMF-based` approach, as it was taking too long to terminate even for small graphs with 300 nodes and 1.5 K edges. The results are similar for other graph sizes that,

---

[4]As a reminder, egonet of a node is the induced subgraph of its neighbors.
[5]The degree of an egonet is defined as the number of incoming and outgoing edges of the subgraph, when viewed as a super-node.

for simplicity, are not shown in the figure. For graphs with 200 nodes and ~ 1.1 K edges (which is the biggest graph for which all the methods were able to terminate), UNI-ALIGN is 1.75× faster than Umeyama's approach; 2× faster than NetAlign-deg; 2, 927× faster than NetAlign-full; and 31,709× faster than the NMF-based approach.

6.5 DISCUSSION

The experiments show that BiG-ALIGN efficiently solves a problem that has been neglected in the literature: the alignment of bipartite graphs. Given that all the efforts have been targeted at aligning uni-partite graphs, why does matching bipartite graphs deserve being studied separately? First, bipartite networks are omnipresent: users like web pages, belong to online communities, access shared files in companies, post in blogs, co-author papers, attend conferences, etc. All these settings can be modeled as bipartite graphs. Second, although it is possible to turn them into unipartite and apply an off-the-shelf algorithm, as shown in the experiments, knowledge of the specific structural characteristics can prove useful in achieving alignments of better quality. Last, this problem enables emerging applications. For instance, one may be able to link the clustering results from different networks by applying soft clustering on the input graphs, and subsequently our method on the obtained node-cluster membership graphs.

Although the main focus of our work is bipartite graph alignment, the latter inspires an alternative way of matching unipartite graphs, by turning them into bipartite. Therefore, we show how our framework can handle any type of input graphs, without any restrictions on their structure. Moreover, it can be applied even to align clouds of points; we can extract features from the points, create a point-to-feature graph, and apply BiG-ALIGN to the latter.

Finally, is our approach simply gradient descent? The answer is negative; gradient descent is the *core* of our algorithm, *but* the projection technique, appropriate initialization and choice of the gradient step, as well as careful handling of known graph properties are the critical design choices that make our algorithm successful (as shown in Section 6.4, where we compare our method to simple gradient descent, BiG-ALIGN-Basic).

6.6 RELATED WORK

The graph alignment problem is of such great interest that there are more than 150 publications proposing different solutions for it, spanning numerous research fields: from data mining to security and re-identification [94, 157], bioinformatics [24, 116, 196], databases [151], chemistry [197], vision, and pattern recognition [49]. Among the suggested approaches are genetic, spectral, clustering algorithms [20, 177], decision trees, expecation-maximization [143], graph edit distance [182], simplex [11], nonlinear optimization [81], iterative HITS-inspired [27, 231], and probabilistic [187]. Some methods that are more efficient for large graphs include a distributed, belief-propagation-based method for protein alignment [34], another message-passing algorithm for aligning sparse networks when some possible matchings are given [23].

We note that all these works are designed for unipartite graphs, while we focus on bipartite graphs.

One of the well-known approaches is Umeyama's near-optimum solution for nearly-isomorphic graphs [212]. The graph matching or alignment problem is formulated as the optimization problem

$$\min_{\mathbf{P}} ||\mathbf{P}\mathbf{A}_1\mathbf{P}^T - \mathbf{A}_2||,$$

where $\mathbf{P}$ is a permutation matrix. The method solves the problem based on the eigendecompositions of the matrices. For symmetric $n \times n$ matrices $\mathbf{A}_1$ and $\mathbf{A}_2$, their eigendecompositions are given by $\mathbf{A}_1 = \mathbf{U}_A\mathbf{\Lambda}_A\mathbf{U}_A{}^T$ and $\mathbf{A}_2 = \mathbf{U}_B\mathbf{\Lambda}_B\mathbf{U}_B{}^T$, where $\mathbf{U}_A$ ($\mathbf{U}_B$) is an orthonormal[6] matrix whose i^{th} column is the eigenvector $\mathbf{v_i}$ of $\mathbf{A}_1$ ($\mathbf{A}_2$), and $\mathbf{\Lambda}_A$ ($\mathbf{\Lambda}_B$) is the diagonal matrix with the corresponding eigenvalues. When $\mathbf{A}_1$ and $\mathbf{A}_2$ are isomorphic, the optimum permutation matrix is obtained by applying the Hungarian algorithm [166] to the matrix $\mathbf{U}_B\mathbf{U}_A{}^T$. This solution is good only if the matrices are isomorphic or nearly isomorphic. Umeyama's approach operates on unipartite, weighted graphs with the *same* number of nodes. Follow-up works employ different constraints for matrix $\mathbf{P}$. For example, the constraint that $\mathbf{P}$ is a doubly stochastic matrix is imposed in [216] and [232], where the proposed formulation, PATH, is based on convex and concave relaxations.

Ding et al. [56] proposed a Non-Negative Matrix Factorization (NMF) approach, which starts from Umeyama's solution, and then applies an iterative algorithm to find the orthogonal matrix $\mathbf{P}$ with the node correspondences. The multiplicative update algorithm for weighted, undirected graphs is:

$$P_{ij} \leftarrow P_{ij}\sqrt{\frac{(\mathbf{A}_1\mathbf{P}\mathbf{A}_2)_{ij}}{(\mathbf{P}\alpha)_{ij}}}$$

and

$$\alpha = \frac{\mathbf{P}^T\mathbf{A}_1\mathbf{P}\mathbf{A}_2 + (\mathbf{P}^T\mathbf{A}_1\mathbf{P}\mathbf{A}_2)^T}{2}.$$

The algorithm stops when convergence is achieved. At that point, $\mathbf{P}$ often has entries that are not in $\{0, 1\}$, so the authors propose applying the Hungarian algorithm for the bipartite graph matching (where "graph mathching" refers to the graph theoretical problem that is a special case of the network flow problem. It does not refer to graph alignment.). The runtime complexity of the NMF-based algorithm is cubic on the number of nodes.

Bradde et al. [34] proposed distributed, heuristic, message-passing algorithms, based on Belief Propagation [228], for protein alignment and prediction of interacting proteins. Independently, Bayati et al. [23] formulated graph matching as an integer quadratic problem, and also proposed message-passing algorithms for aligning sparse networks. In addition to the input matrices $\mathbf{A}_1$ and $\mathbf{A}_2$, a sparse and weighted bipartite graph $\mathbf{L}$ between the vertices of $\mathbf{A}_1$ and $\mathbf{A}_2$ is also needed. The edges of graph $\mathbf{L}$ represent the possible node matchings between the two

---

[6]A matrix $\mathbf{R}$ is orthonormal if it is a square matrix with real entries such that $\mathbf{R}^T\mathbf{R} = \mathbf{R}\mathbf{R}^T = I$.

graphs and their weights capture the similarity between the connected nodes). Singh et al. [196] had proposed the use of the full bipartite graph earlier. However, as we have shown in our experiments, this variation has high memory requirements and does not scale well for large graphs. A related problem formulation was studied by Klau [116], who proposed a Lagrangian relaxation approach combined with branch-and-bound to align protein-protein interaction networks and classify metabolic subnetworks.

In all these works, the graphs that are studied are unipartite, while, in this chapter, we focus on bipartite graphs, and extend BIG-ALIGN to handle unipartite graphs.

CHAPTER 7

Conclusions and Further Research Problems

Graphs are very powerful representations of data and the relations among them. The Web, friendships and communications, collaborations and phone calls, traffic flow, or brain functions are only few examples of the processes that are naturally captured by graphs, which often span *hundreds of millions* or *billions* of nodes and edges. Within this abundance of interconnected data, a key challenge is the extraction of *useful knowledge* in a *scalable* way.

In this book, we focused on fast and principled methods for the individual and collective graph exploration and analysis in order to gain insights into the underlying data and phenomena. The book focused on exploring large amounts of data via scalable (1) summarization techniques, which aim to provide a compact and interpretable representation of static or dynamic graphs, and node behaviors, and (2) techniques that are based on similarity or affinity (e.g., via guilt-by-association methods, similarity at the node or graph level, and alignment). These methods are powerful, can be applied in various settings where the data can be represented as a graph, and have applications ranging from anomaly detection in static and dynamic graphs (e.g., email communications or computer network monitoring), to clustering and classification, to re-identification across networks and visualization.

Next, we outline some of the challenging research directions stemming in the space of automating the sense-making of large, real-world networks.

Summarization for Complex Data and Novel Definitions   In this book, we reviewed a new, alternative way of summarizing large undirected and unweighted graphs. Although many approaches tackle the case of plain static networks, most real-world graphs, such as social and communication networks, rapidly evolve over time. For example, Twitter can be modeled as a time-evolving network that includes edges for follow, retweet, and messaging activities. New methods for summarizing more complex data—such as streaming graph data, multi-layer or spatiotermporal networks, and time-evolving networks with side information—have the potential for significant impact in various domains. Moreover, hierarchical, interactive, or domain-aware summarization is a promising direction that can transform the way analysts and scientists interact with their data.

Unifying Summarization and Visualization of Large Graphs   The continuous generation of interconnected data creates the need for summarizing them to extract easy-to-understand information. We claim that visualization is the other side of the same coin; summarization aims at minimizing the number of bits, and visualization the number of pixels. Currently, visualization of large graphs is almost impossible. It is based on global structures and yields an uninformative clutter of nodes and edges. Formalizing graph visualization by unifying it with information theory might be beneficial for making sense of large amounts of networked data.

Evaluation Techniques   One of the main challenges in summarization is the lack of standard evaluation techniques. Currently, summarization methods are being evaluated within their application domain: For instance, compression-based methods are evaluated in terms of bit minimization (although the best possible compression is not their sole goal), while query-preserving methods are evaluated in terms of query speedup and accuracy. Having some common evaluation metrics, in addition to the application-specific ones, would be useful for comparing new approaches to existing ones.

Representation Learning   Node representation learning is a new and popular area in data mining [84, 86, 175, 202, 218], which has focused more on concrete tasks, such as link prediction and classification within one network. A new, promising direction is to go beyond factorization-based latent representations and explore graph summarization techniques based on deep node representations that are learned automatically from the context encoded in the graph. Moreover, extending node representation learning techniques to handle multiple graphs collectively has not been studied yet, but it can have significant impact on network similarity, alignments, and other complex tasks that go beyond one network.

Multi-source Graph Mining   The second part of this book covered scalable algorithms for collectively mining multiple graphs, with focus on summarization of temporal networks, similarity between pairs of graphs, and pairwise network alignment. A natural extension is to consider broader and more complex problems on multiple graphs that are collected from multiple sources, which are characterized by a series of challenges: (i) Structural Heterogeneity: The observed or inferred graphs may have largely different number of entities with no or small overlap, or alignment between them may be missing, partially observed or very noisy; (ii) Temporal Heterogeneity: Time-evolving graphs may be observed at different granularities (e.g., computer networks might be observed per minute, while user interactions over these networks might be aggregated at an hourly or daily rate); and (iii) Noise: Although combining different data sources and obtaining a composite picture of the data can give insights that would be missed when analyzing each graph in isolation, the aggregated noise from these sources can degrade the performance of graph mining tasks. Developing robust and scalable methods has the potential to change the way multi-source graphs are being analyzed and to lead to new problem definitions.

Big Data Systems for Scalability   Scalability will continue to be an integral part of real-world applications. HADOOP is appropriate for several tasks, but falls short when there is need for real-time analytics, iterative approaches, and specific structures. Moving forward, it would be beneficial to explore other big data systems (e.g., Spark, GraphLab, Cloudera Impala) that match the intrinsic nature of the data at hand and special requirements of the methods. Moreover, investigating how to exploit the capabilities of big data systems to provide the analyst with fast, approximate answers, which will then be refined further depending on the time constraints, can contribute to even more scalable approaches to sift through and understand the ever-growing amounts of interconnected data.

Applications to Other Domains It is natural to align graph mining techniques with the type of data, and by extension, the application at hand. Graph mining has traditionally been applied in many domains, including social and behavioral sciences, genomics, language understanding, and more. In this book, we presented analysis of brain graphs and several scientific discoveries, which constitute just an initial step in the realm of analyzing brain networks to understand how the brain works. Many questions about the function of the brain and mental diseases remain unanswered, despite the huge economic cost of neurological disorders. Scalable and accurate graph-based computational methods can contribute to the brain research, which is supported by the U.S. government through the Brain Research through Advancing Innovative Neurotechnologies (BRAIN) Initiative. Moreover, applications in IoT (Internet of Things) and autonomous connected vehicles in transportation engineering, which require summarization techniques, are only a few of the new frontiers to explore in the future.

In conclusion, understanding, mining, and managing large graphs have numerous high-impact applications and research challenges.

Bibliography

[1] B. Adhikari, Y. Zhang, A. Bharadwaj, and A. Prakash. Condensing temporal networks using propagation. In *Proc. of the 17th SIAM International Conference on Data Mining (SDM)*, Houston, TX, 2017. DOI: 10.1137/1.9781611974973.47. 95

[2] C. Aggarwal and K. Subbian. Evolutionary network analysis: A survey. *ACM Computing Surveys*, 47(1):10:1–10:36, May 2014. DOI: 10.1145/2601412. 75

[3] C. C. Aggarwal and S. Y. Philip. Online analysis of community evolution in data streams. In *Proc. of the 5th SIAM International Conference on Data Mining (SDM)*, Newport Beach, CA, 2005. DOI: 10.1137/1.9781611972757.6. 93

[4] E. M. Airoldi, D. M. Blei, S. E. Fienberg, and E. P. Xing. Mixed membership stochastic blockmodels. *Journal of Machine Learning Research*, 9:1981–2014, 2008. 22, 47

[5] L. Akoglu*, D. H. Chau*, U. Kang*, D. Koutra*, and C. Faloutsos. OPAvion: Mining and visualization in large graphs. In *Proc. of the ACM International Conference on Management of Data (SIGMOD)*, pages 717–720, Scottsdale, AZ, 2012. DOI: 10.1145/2213836.2213941. 47

[6] L. Akoglu and C. Faloutsos. Event detection in time series of mobile communication graphs. In *27th Army Science Conference*, 2010. 141

[7] L. Akoglu, M. McGlohon, and C. Faloutsos. OddBall: Spotting anomalies in weighted graphs. In *Proc. of the 14th Pacific-Asia Conference on Knowledge Discovery and Data Mining (PAKDD)*, Hyderabad, India, 2010. DOI: 10.1007/978-3-642-13672-6_40. 47

[8] L. Akoglu, H. Tong, and D. Koutra. Graph-based anomaly detection and description: A survey. *Data Mining and Knowledge Discovery (DAMI)*, April 2014. DOI: 10.1007/s10618-014-0365-y. 141

[9] R. Albert and A.-L. Barabási. Statistical mechanics of complex networks. *CoRR*, cond-mat/0106096, 2001. DOI: 10.1103/revmodphys.74.47. 150

[10] D. Aldous and J. A. Fill. Reversible Markov chains and random walks on graphs, 2002. Unfinished monograph, recompiled 2014. http://www.stat.berkeley.edu/$\sim$aldous/RWG/book.html 99

[11] H. A. Almohamad and S. O. Duffuaa. A linear programming approach for the weighted graph matching problem. *IEEE Transactions on Pattern Analysis and Machine Intelligence*, 15(5):522–525, 1993. DOI: 10.1109/34.211474. 163

[12] B. Alper, B. Bach, N. Henry Riche, T. Isenberg, and J.-D. Fekete. Weighted graph comparison techniques for brain connectivity analysis. In *Proc. of the SIGCHI Conference on Human Factors in Computing Systems, (CHI'13)*, pages 483–492, New York, ACM, 2013. DOI: 10.1145/2470654.2470724. 139

[13] C. J. Alpert, A. B. Kahng, and S.-Z. Yao. Spectral partitioning with multiple eigenvectors. *Discrete Applied Mathematics*, 90(1):3–26, 1999. DOI: 10.1016/s0166-218x(98)00083-3. 83, 93

[14] R. Andersen, F. Chung, and K. Lang. Local graph partitioning using PageRank vectors. In *Proc. of the 47th Annual IEEE Symposium on Foundations of Computer Science*, pages 475–486, IEEE Computer Society, 2006. DOI: 10.1109/focs.2006.44. 50

[15] K. Andrews, M. Wohlfahrt, and G. Wurzinger. Visual graph comparison. In *13th International Conference on Information Visualization—Showcase (IV)*, pages 62–67, July 2009. DOI: 10.1109/iv.2009.108. 139

[16] A. Apostolico and G. Drovandi. Graph compression by BFS. *Algorithms*, 2(3):1031–1044, 2009. DOI: 10.3390/a2031031. 46

[17] M. Araujo, S. Günnemann, G. Mateos, and C. Faloutsos. Beyond blocks: Hyperbolic community detection. In *Proc. of the European Conference on Machine Learning and Principles and Practice of Knowledge Discovery in Databases (ECML PKDD)*, pages 50–65, Nancy, France, Springer, 2014. DOI: 10.1007/978-3-662-44848-9_4. 45

[18] M. Araujo, S. Papadimitriou, S. Günnemann, C. Faloutsos, P. Basu, A. Swami, E. E. Papalexakis, and D. Koutra. Com2: Fast automatic discovery of temporal ("Comet") communities. In *Proc. of the 18th Pacific-Asia Conference on Knowledge Discovery and Data Mining (PAKDD)*, pages 271–283, Tainan, Taiwan, Springer, 2014. DOI: 10.1007/978-3-319-06605-9_23. 77, 93

[19] AS-Oregon dataset. http://topology.eecs.umich.edu/data.html

[20] X. Bai, H. Yu, and E. R. Hancock. Graph matching using spectral embedding and alignment. In *Proc. of the 17th International Conference on Pattern Recognition*, volume 3, pages 398–401, August 2004. DOI: 10.1109/icpr.2004.1334550. 163

[21] N. Barbieri, F. Bonchi, and G. Manco. Cascade-based community detection. In *WSDM*, pages 33–42, 2013. DOI: 10.1145/2433396.2433403. 75, 93

[22] M. Bayati, M. Gerritsen, D. Gleich, A. Saberi, and Y. Wang. Algorithms for large, sparse network alignment problems. In *Proc. of the 9th IEEE International Conference on Data Mining (ICDM)*, pages 705–710, Miami, FL, 2009. DOI: 10.1109/icdm.2009.135. 143

[23] M. Bayati, D. F. Gleich, A. Saberi, and Y. Wang. Message-passing algorithms for sparse network alignment. *ACM Transactions on Knowledge Discovery from Data*, 7(1):3:1–3:31, Helen Martin, 2013. DOI: 10.1145/2435209.2435212. 140, 143, 157, 163, 164

[24] J. Berg and M. Lässig. Local graph alignment and motif search in biological networks. *Proc. of the National Academy of Sciences*, 101(41):14689–14694, October 2004. DOI: 10.1073/pnas.0305199101. 163

[25] M. Berlingerio, D. Koutra, T. Eliassi-Rad, and C. Faloutsos. Network similarity via multiple social theories. *IEEE/ACM Conference on Advances in Social Networks Analysis and Mining (ASONAM'13)*, 2013. DOI: 10.1145/2492517.2492582. 140

[26] E. Bertini and G. Santucci. By chance is not enough: Preserving relative density through non uniform sampling. In *Proc. of the Information Visualisation*, 2004. DOI: 10.1109/iv.2004.1320207. 48

[27] V. D. Blondel, A. Gajardo, M. Heymans, P. Senellart, and P. V. Dooren. A measure of similarity between graph vertices: Applications to synonym extraction and Web searching. *SIAM Review*, 46(4):647–666, April 2004. DOI: 10.1137/s0036144502415960. 163

[28] V. D. Blondel, J.-L. Guillaume, R. Lambiotte, and E. Lefebvre. Fast unfolding of communities in large networks. *Journal of Statistical Mechanics: Theory and Experiment*, 2008(10):P10008, 2008. DOI: 10.1088/1742-5468/2008/10/p10008. 93

[29] A. Blum and S. Chawla. Learning from labeled and unlabeled data using graph mincuts. In *Proc. of the 18th International Conference on Machine Learning*, pages 19–26, Morgan Kaufmann, San Francisco, CA, 2001. 51

[30] S. Boccaletti, G. Bianconi, R. Criado, C. I. Del Genio, J. Gómez-Gardenes, M. Romance, I. Sendina-Nadal, Z. Wang, and M. Zanin. The structure and dynamics of multilayer networks. *Physics Reports*, 544(1):1–122, 2014. DOI: 10.1016/j.physrep.2014.07.001. 93

[31] P. Boldi and S. Vigna. The webgraph framework I: Compression techniques. In *Proc. of the 13th International Conference on World Wide Web (WWW)*, New York, 2004. DOI: 10.1145/988672.988752. 46

[32] K. M. Borgwardt and H.-P. Kriegel. Shortest-path Kernels on graphs. In *Proc. of the 5th IEEE International Conference on Data Mining (ICDM'05)*, pages 74–81, IEEE Computer Society, Washington, DC, 2005. DOI: 10.1109/icdm.2005.132. 140

[33] S. Boyd and L. Vandenberghe. *Convex Optimization*. Cambridge University Press, New York, 2004. DOI: 10.1017/cbo9780511804441. 151

[34] S. Bradde, A. Braunstein, H. Mahmoudi, F. Tria, M. Weigt, and R. Zecchina. Aligning graphs and finding substructures by a cavity approach. *Europhysics Letters*, 89, 2010. DOI: 10.1209/0295-5075/89/37009. 140, 143, 163, 164

[35] S. Brin and L. Page. The anatomy of a large-scale hypertextual web search engine. *Computer Networks and ISDN Systems*, 30(1-7):107–117, 1998. DOI: 10.1016/s0169-7552(98)00110-x. 11, 50, 53, 99, 140

[36] A. Broder, R. Kumar, F. Maghoul, P. Raghavan, S. Rajagopalan, R. Stata, A. Tomkins, and J. Wiener. Graph structure in the Web. *Computer Network*, 33(1-6):309–320, June 2000. DOI: 10.1016/s1389-1286(00)00083-9. 150

[37] C. Budak, D. Agrawal, and A. El Abbadi. Diffusion of information in social networks: Is it all local? In *ICDM*, pages 121–130, 2012. DOI: 10.1109/icdm.2012.74. 75

[38] H. Bunke, P. J. Dickinson, M. Kraetzl, and W. D. Wallis. *A Graph-theoretic Approach to Enterprise Network Dynamics (PCS)*. Birkhauser, 2006. DOI: 10.1007/978-0-8176-4519-9. 117, 138, 139

[39] R. S. Caceres, T. Y. Berger-Wolf, and R. Grossman. Temporal scale of processes in dynamic networks. In *Proc. of the Data Mining Workshops (ICDMW) at the 11th IEEE International Conference on Data Mining (ICDM)*, pages 925–932, Vancouver, Canada, 2011. DOI: 10.1109/icdmw.2011.165. 97

[40] D. Chakrabarti, S. Papadimitriou, D. S. Modha, and C. Faloutsos. Fully automatic cross-associations. In *Proc. of the 10th ACM International Conference on Knowledge Discovery and Data Mining (SIGKDD)*, pages 79–88, Seattle, WA, 2004. DOI: 10.1145/1014052.1014064. 47, 83, 93

[41] D. Chakrabarti, Y. Zhan, D. Blandford, C. Faloutsos, and G. Blelloch. NetMine: New mining tools for large graphs. In *SIAM-data Mining Workshop on Link Analysis, Counterterrorism and Privacy*, 2004. 17, 26

[42] D. H. Chau, A. Kittur, J. I. Hong, and C. Faloutsos. Apolo: Making sense of large network data by combining rich user interaction and machine learning. In *Proc. of the 17th ACM International Conference on Knowledge Discovery and Data Mining (SIGKDD)*, San Diego, CA, 2011. DOI: 10.1145/1978942.1978967. 47

[43] D. H. Chau, C. Nachenberg, J. Wilhelm, A. Wright, and C. Faloutsos. Large scale graph mining and inference for Malware detection. In *Proc. of the 11th SIAM International Conference on Data Mining (SDM)*, pages 131–142, Mesa, AZ, 2011. DOI: 10.1137/1.9781611972818.12. 52, 53, 141

[44] A. Chechetka and C. E. Guestrin. Focused belief propagation for query-specific inference. In *International Conference on Artificial Intelligence and Statistics (AISTATS)*, May 2010. 52

[45] Y. Chen, E. K. Garcia, M. R. Gupta, A. Rahimi, and L. Cazzanti. Similarity-based classification: Concepts and algorithms. *Journal of Machine Learning Research*, 10:747–776, June 2009. 97

[46] F. Chierichetti, R. Kumar, S. Lattanzi, M. Mitzenmacher, A. Panconesi, and P. Raghavan. On compressing social networks. In *Proc. of the 15th ACM International Conference on Knowledge Discovery and Data Mining (SIGKDD)*, pages 219–228, Paris, France, 2009. DOI: 10.1145/1557019.1557049. 46

[47] N. A. Christakis and J. H. Fowler. The spread of obesity in a large social network over 32 years. *New England Journal of Medicine*, 357(4):370–379, 2007. DOI: 10.1056/nejmsa066082. 49

[48] W. W. Cohen. Graph walks and graphical models. Technical Report CMU-ML-10-102, Carnegie Mellon University, March 2010. 55

[49] D. Conte, P. Foggia, C. Sansone, and M. Vento. Thirty years of graph matching in pattern recognition. *International Journal of Pattern Recognition and Artificial Intelligence*, 18(3):265–298, 2004. DOI: 10.1142/s0218001404003228. 140, 143, 163

[50] D. J. Cook and L. B. Holder. Substructure discovery using minimum description length and background knowledge. *Journal of Artificial Intelligence Research*, 1:231–255, 1994. 17, 26, 46, 93

[51] F. Costa and K. De Grave. Fast neighborhood subgraph pairwise distance Kernel. In *Proc. of the 26th International Conference on Machine Learning*, 2010. 140

[52] T. M. Cover and J. A. Thomas. *Elements of Information Theory*. Wiley-Interscience, New York, 2006. DOI: 10.1002/0471200611. 23, 80

[53] DBLP network dataset. `konect.uni-koblenz.de/networks/dblp_coauthor`, July 2014.

[54] I. Dhillon, Y. Guan, and B. Kulis. A fast Kernel-based multilevel algorithm for graph clustering. In *Proc. of the 11th ACM International Conference on Knowledge Discovery and Data Mining (SIGKDD)*, pages 629–634, Chicago, IL, ACM, 2005. DOI: 10.1145/1081870.1081948. 83, 93

[55] I. S. Dhillon, S. Mallela, and D. S. Modha. Information-theoretic co-clustering. In *Proc. of the 9th ACM International Conference on Knowledge Discovery and Data Mining (SIGKDD)*, pages 89–98, Washington, DC, 2003. DOI: 10.1145/956750.956764. 93

[56] C. H. Q. Ding, T. Li, and M. I. Jordan. Nonnegative matrix factorization for combinatorial optimization: spectral clustering, graph matching, and clique finding. In *Proc. of the 8th IEEE International Conference on Data Mining (ICDM)*, pages 183–192, Pisa, Italy, 2008. DOI: 10.1109/icdm.2008.130. 144, 145, 150, 157, 164

[57] P. Doyle and J. L. Snell. *Random Walks and Electric Networks*, volume 22. Mathematical Association America, New York, 1984. DOI: 10.5948/upo9781614440222. 50, 99, 140

[58] C. Dunne and B. Shneiderman. Motif simplification: Improving network visualization readability with fan, connector, and clique glyphs. In *Proc. of the SIGCHI Conference on Human Factors in Computing Systems (CHI)*, pages 3247–3256, ACM, 2013. DOI: 10.1145/2470654.2466444. 47

[59] H. Elghawalby and E. R. Hancock. Measuring graph similarity using spectral geometry. In *Proc. of the 5th International Conference on Image Analysis and Recognition (ICIAR)*, pages 517–526, 2008. DOI: 10.1007/978-3-540-69812-8_51. 140

[60] Enron dataset. http://www.cs.cmu.edu/~enron

[61] C. Erten, P. J. Harding, S. G. Kobourov, K. Wampler, and G. Yee. GraphAEL: Graph animations with evolving layouts. In *Proc. of the 11th International Symposium in Graph Drawing (GD)*, volume 2912, pages 98–110, Perugia, Italy, 2003. DOI: 10.1007/978-3-540-24595-7_9. 139

[62] D. Eswaran, S. Günnemann, C. Faloutsos, D. Makhija, and M. Kumar. Zoobp: Belief propagation for heterogeneous networks. *Proc. of the VLDB Endowment*, 10(5):625–636, January 2017. DOI: 10.14778/3055540.3055554. 67

[63] C. Faloutsos and V. Megalooikonomou. On data mining, compression and Kolmogorov complexity. In *Data Mining and Knowledge Discovery*, volume 15, pages 3–20, Springer-Verlag, 2007. DOI: 10.1007/s10618-006-0057-3. 46

[64] M. Faloutsos, P. Faloutsos, and C. Faloutsos. On power-law relationships of the internet topology. *ACM SIGCOMM Computer Communication Review*, 29(4):251–262, August 1999. DOI: 10.1145/316194.316229. 17, 150

[65] K. Faust and S. Wasserman. Blockmodels: Interpretation and evaluation. *Social Networks*, 14(1-2):5–61, 1992. DOI: 10.1016/0378-8733(92)90013-w. 47

[66] P. F. Felzenszwalb and D. P. Huttenlocher. Efficient belief propagation for early vision. *International Journal of Computer Vision*, 70(1):41–54, 2006. DOI: 10.1109/cvpr.2004.1315041. 52

[67] J. Feng, X. He, N. Hubig, C. Böhm, and C. Plant. Compression-based graph mining exploiting structure primitives. In *Proc. of the 14th IEEE International Conference on Data Mining (ICDM)*, pages 181–190, Dallas, TX, IEEE, 2013. DOI: 10.1109/icdm.2013.56. 46

[68] J. Ferlez, C. Faloutsos, J. Leskovec, D. Mladenic, and M. Grobelnik. Monitoring network evolution using MDL. *Proc. of the 24th International Conference on Data Engineering (ICDE)*, pages 1328–1330, Cancun, Mexico, 2008. DOI: 10.1109/icde.2008.4497545. 94

[69] M. Fiedler. Algebraic connectivity of graphs. *Czechoslovak Mathematical Journal*, 23(98):298–305, 1973. 139

[70] Flickr. http://www.flickr.com

[71] D. Fogaras and B. Rácz. Towards scaling fully personalized pagerank. In *Algorithms and Models for the Web-graph*, volume 3243 of *Lecture Notes in Computer Science*, pages 105–117, 2004. DOI: 10.1007/978-3-540-30216-2_9. 50

[72] J. H. Fowler and N. A. Christakis. Dynamic spread of happiness in a large social network: Longitudinal analysis over 20 years in the Framingham heart study. *British Medical Journal*, 2008. DOI: 10.1136/bmj.a2338. 49

[73] L. C. Freeman. A set of measures of centrality based on betweenness. *Sociometry*, pages 35–41, 1977. DOI: 10.2307/3033543. 141

[74] W. Fu, L. Song, and E. P. Xing. Dynamic mixed membership blockmodel for evolving networks. In *Proc. of the 26th Annual International Conference on Machine Learning, (ICML'09)*, pages 329–336, New York, ACM, 2009. DOI: 10.1145/1553374.1553416. 94

[75] K. Fukunaga. *Introduction to Statistical Pattern Recognition*. Access online via Elsevier, 1990. 114

[76] J. Gao, F. Liang, W. Fan, Y. Sun, and J. Han. Graph-based consensus maximization among multiple supervised and unsupervised models. In *Proc. of the 23rd Annual Conference on Neural Information Processing Systems (NIPS)*, Whistler, Canada, 2009. DOI: 10.1109/tkde.2011.206. 68

[77] T. Gärtner, P. A. Flach, and S. Wrobel. On graph Kernels: Hardness results and efficient alternatives. In *Proc. of the 16th Annual Conference on Computational Learning Theory and the 7th Kernel Workshop*, 2003. DOI: 10.1007/978-3-540-45167-9_11. 140

[78] W. Gatterbauer, S. Günnemann, D. Koutra, and C. Faloutsos. Linearized and single-pass belief propagation. *Proc. of the VLDB Endowment*, 8(5):581–592, 2015. DOI: 10.14778/2735479.2735490. 3, 65, 67

[79] D. F. Gleich and M. W. Mahoney. Using local spectral methods to robustify graph-based learning algorithms. In *Proc. of the 21th ACM SIGKDD International Conference on Knowledge Discovery and Data Mining*, pages 359–368, ACM, 2015. DOI: 10.1145/2783258.2783376. 57

[80] M. Gleicher, D. Albers Szafir, R. Walker, I. Jusufi, C. D. Hansen, and J. C. Roberts. Visual comparison for information visualization. *Information Visualization*, 10(4):289–309, October 2011. DOI: 10.1177/1473871611416549. 139

[81] S. Gold and A. Rangarajan. A graduated assignment algorithm for graph matching. *IEEE Transactions on Pattern Analysis and Machine Intelligence*, 18(4):377–388, 1996. DOI: 10.1109/34.491619. 163

[82] M. Gomez Rodriguez, J. Leskovec, and A. Krause. Inferring networks of diffusion and influence. In *Proc. of the 16th ACM International Conference on Knowledge Discovery and Data Mining (SIGKDD)*, pages 1019–1028, Washington, DC, ACM, 2010. DOI: 10.1145/1835804.1835933. 75

[83] J. Gonzalez, Y. Low, and C. Guestrin. Residual splash for optimally parallelizing belief propagation. *Journal of Machine Learning Research—Proceedings Track*, 5:177–184, 2009. 52

[84] P. Goyal and E. Ferrara. Graph embedding techniques, applications, and performance: A survey. *arXiv preprint arXiv:1705.02801*, 2017. 168

[85] W. R. Gray, J. A. Bogovic, J. T. Vogelstein, B. A. Landman, J. L. Prince, and R. J. Vogelstein. Magnetic resonance connectome automated pipeline: An overview. *Pulse, IEEE*, 3(2):42–48, 2012. DOI: 10.1109/mpul.2011.2181023. 134

[86] A. Grover and J. Leskovec. Node2vec: Scalable feature learning for networks. *ACM*, 2016. DOI: 10.1145/2939672.2939754. 168

[87] D. Gruhl, R. V. Guha, D. Liben-Nowell, and A. Tomkins. Information diffusion through blogspace. In *World Wide Web Conference*, pages 491–501, New York, May 2004. DOI: 10.1145/988672.988739. 75, 93

[88] P. Grünwald. *The Minimum Description Length Principle*. MIT Press, 2007. 20, 44, 46

[89] P. D. Grünwald. *The Minimum Description Length Principle (Adaptive Computation and Machine Learning)*. The MIT Press, 2007. 21

[90] R. Guha, R. Kumar, P. Raghavan, and A. Tomkins. Propagation of trust and distrust. In *Proc. of the 13th International Conference on World Wide Web (WWW)*, pages 403–412, New York, ACM, 2004. DOI: 10.1145/988672.988727.

[91] Hadoop information. http://hadoop.apache.org/ 52

[92] M. Hascoët and P. Dragicevic. Interactive graph matching and visual comparison of graphs and clustered graphs. In *Proc. of the International Working Conference on Advanced Visual Interfaces, (AVI'12)*, pages 522–529, New York, ACM, 2012. DOI: 10.1145/2254556.2254654. 139

[93] T. H. Haveliwala. Topic-sensitive PageRank: A context-sensitive ranking algorithm for Web search. *IEEE Transactions on Knowledge and Data Engineering*, 15(4):784–796, 2003. DOI: 10.1109/tkde.2003.1208999. 50, 55, 99, 140

[94] K. Henderson, B. Gallagher, L. Li, L. Akoglu, T. Eliassi-Rad, H. Tong, and C. Faloutsos. It's who you know: Graph mining using recursive structural features. In *Proc. of the 17th ACM International Conference on Knowledge Discovery and Data Mining (SIGKDD)*, pages 663–671, San Diego, CA, 2011. DOI: 10.1145/2020408.2020512. 163

[95] T. Horváth, T. Gärtner, and S. Wrobel. Cyclic pattern Kernels for predictive graph mining. In *Proc. of the 10th ACM International Conference on Knowledge Discovery and Data Mining (SIGKDD)*, pages 158–167, Seattle, WA, ACM, 2004. DOI: 10.1145/1014052.1014072. 140

[96] C. Hübler, H.-P. Kriegel, K. Borgwardt, and Z. Ghahramani. Metropolis algorithms for representative subgraph sampling. In *Proc. of the 8th IEEE International Conference on Data Mining, (ICDM'08)*, pages 283–292, Washington, DC, IEEE Computer Society, 2008. DOI: 10.1109/icdm.2008.124. 46

[97] Y. Hulovatyy, H. Chen, and T. Milenković. Exploring the structure and function of temporal networks with dynamic graphlets. *Bioinformatics*, 31(12):i171–i180, 2015. DOI: 10.1093/bioinformatics/btv227. 94

[98] G. Jeh and J. Widom. SimRank: A measure of structural-context similarity. In *Proc. of the 8th ACM International Conference on Knowledge Discovery and Data Mining (SIGKDD)*, pages 538–543, Edmonton, Alberta, Canada, ACM, 2002. DOI: 10.1145/775107.775126. 50, 140

[99] M. Ji, Y. Sun, M. Danilevsky, J. Han, and J. Gao. Graph regularized transductive classification on heterogeneous information networks. In *Proc. of the European Conference on Machine Learning and Principles and Practice of Knowledge Discovery in Databases (ECML PKDD)*, pages 570–586, Barcelona, Spain, 2010. DOI: 10.1007/978-3-642-15880-3_42. 51, 53

[100] R. Jin, C. Wang, D. Polshakov, S. Parthasarathy, and G. Agrawal. Discovering frequent topological structures from graph datasets. In *Proc. of the 11th ACM International Conference on Knowledge Discovery and Data Mining (SIGKDD)*, pages 606–611, Chicago, IL, 2005. DOI: 10.1145/1081870.1081944. 93

[101] T. Kamada and S. Kawai. An algorithm for drawing general undirected graphs. *Information Processing Letters*, 31:7–15, 1989. DOI: 10.1016/0020-0190(89)90102-6. 1, 19

[102] U. Kang, D. H. Chau, and C. Faloutsos. Mining large graphs: Algorithms, inference, and discoveries. In *Proc. of the 27th International Conference on Data Engineering (ICDE)*, pages 243–254, Hannover, Germany, 2011. DOI: 10.1109/icde.2011.5767883. 71

[103] U. Kang and C. Faloutsos. Beyond "Caveman Communities": Hubs and spokes for graph compression and mining. In *Proc. of the 11th IEEE International Conference on Data Mining (ICDM)*, Vancouver, Canada, 2011. DOI: 10.1109/icdm.2011.26. 83, 88, 94, 153

[104] U. Kang, J.-Y. Lee, D. Koutra, and C. Faloutsos. Net-Ray: Visualizing and mining web-scale graphs. In *Proc. of the 18th Pacific-Asia Conference on Knowledge Discovery and Data Mining (PAKDD)*, Tainan, Taiwan, 2014. DOI: 10.1007/978-3-319-06608-0_29. 48

[105] U. Kang, H. Tong, and J. Sun. Fast random walk graph Kernel. In *Proc. of the 12th SIAM International Conference on Data Mining (SDM)*, Anaheim, CA, 2012. DOI: 10.1137/1.9781611972825.71. 140

[106] U. Kang, C. E. Tsourakakis, and C. Faloutsos. PEGASUS: A peta-scale graph mining system—implementation and observations. *Proc. of the 9th IEEE International Conference on Data Mining (ICDM)*, Miami, FL, 2009. DOI: 10.1109/icdm.2009.14. 70

[107] B. Karrer and M. E. J. Newman. Stochastic blockmodels and community structure in networks. *Physics Review*, E 83, 2011. DOI: 10.1103/physreve.83.016107. 22, 47

[108] G. Karypis and V. Kumar. METIS: Unstructured graph partitioning and sparse matrix ordering system. The University of Minnesota, 2, 1995. 83, 104

[109] G. Karypis and V. Kumar. Multilevel k-way hypergraph partitioning. In *Proc. of the IEEE 36th Conference on Design Automation Conference (DAC)*, pages 343–348, New Orleans, LA, 1999. DOI: 10.1145/309847.309954. 17, 26, 93

[110] H. Kashima, K. Tsuda, and A. Inokuchi. Marginalized Kernels between labeled graphs. In *Proc. of the 20th International Conference on Machine Learning*, pages 321–328, AAAI Press, 2003. 140

[111] L. Katz. A new status index derived from sociometric analysis. *Psychometrika*, 18(1):39–43, March 1953. DOI: 10.1007/bf02289026. 101

[112] A. K. Kelmans. Comparison of graphs by their number of spanning trees. *Discrete Mathematics*, 16(3):241–261, 1976. DOI: 10.1016/0012-365x(76)90102-3. 140

[113] N. S. Ketkar, L. B. Holder, and D. J. Cook. SUBDUE: Compression-based frequent pattern discovery in graph data. *Proc. of the 1st International Workshop on Open Source Data Mining: Frequent Pattern Mining Implementations in Conjunction with the 11th ACM International Conference on Knowledge Discovery and Data Mining (SIGKDD)*, Chicago, IL, August 2005. DOI: 10.1145/1133905.1133915. 83

[114] H.-N. Kim and A. El Saddik. Personalized PageRank vectors for tag recommendations: Inside FolkRank. In *Proc. of the 5th ACM Conference on Recommender Systems*, pages 45–52, 2011. DOI: 10.1145/2043932.2043945. 53, 141

[115] M. Kivelä, A. Arenas, M. Barthelemy, J. P. Gleeson, Y. Moreno, and M. A. Porter. Multilayer networks. *Journal of Complex Networks*, 2(3):203–271, 2014. DOI: 10.1093/comnet/cnu016. 93

[116] G. W. Klau. A new graph-based method for pairwise global network alignment. *BMC Bioinformatics*, 10(S-1), 2009. DOI: 10.1186/1471-2105-10-s1-s59. 163, 165

[117] J. Kleinberg, R. Kumar, P. Raghavan, S. Rajagopalan, and A. Tomkins. The Web as a graph: Measurements, models and methods. In *International Computing and Combinatorics Conference*, Berlin, Germany, Springer, 1999. DOI: 10.1007/3-540-48686-0_1. 17, 21

[118] J. M. Kleinberg. Authoritative sources in a hyperlinked environment. *Journal of the ACM (JACM)*, 46(5):604–632, 1999. DOI: 10.1145/324133.324140. 141

[119] B. Klimt and Y. Yang. Introducing the Enron corpus. In *Proc. of the 1st Conference on E-mail and Anti-spam*, Mountain View, CA, 2004.

[120] A. Koopman and A. Siebes. Discovering relational items sets efficiently. In *Proc. of the 8th SIAM International Conference on Data Mining (SDM)*, pages 108–119, Atlanta, GA, 2008. DOI: 10.1137/1.9781611972788.10. 46

[121] A. Koopman and A. Siebes. Characteristic relational patterns. In *Proc. of the 15th ACM International Conference on Knowledge Discovery and Data Mining (SIGKDD)*, pages 437–446, Paris, France, 2009. DOI: 10.1145/1557019.1557071. 45, 46

[122] Y. Koren, S. C. North, and C. Volinsky. Measuring and extracting proximity in networks. In *Proc. of the 12th ACM International Conference on Knowledge Discovery and Data Mining (SIGKDD)*, Philadelphia, PA, 2006. DOI: 10.1145/1150402.1150432. 50

[123] D. Koutra, U. Kang, J. Vreeken, and C. Faloutsos. VoG: Summarizing and understanding large graphs. In *Proc. of the 14th SIAM International Conference on Data Mining (SDM)*, pages 91–99, Philadelphia, PA, 2014. DOI: 10.1137/1.9781611973440.11. 3, 94

[124] D. Koutra, U. Kang, J. Vreeken, and C. Faloutsos. Summarizing and understanding large graphs. In *Statistical Analysis and Data Mining*. John Wiley & Sons, Inc., 2015. DOI: 10.1002/sam.11267. 3, 94, 153

[125] D. Koutra, T.-Y. Ke, U. Kang, D. H. Chau, H.-K. K. Pao, and C. Faloutsos. Unifying guilt-by-association approaches: Theorems and fast algorithms. In *Proc. of the European Conference on Machine Learning and Principles and Practice of Knowledge Discovery in Databases (ECML PKDD)*, pages 245–260, Athens, Greece, 2011. DOI: 10.1007/978-3-642-23783-6_16. 3, 4, 27, 99, 128, 141

[126] D. Koutra, N. Shah, J. Vogelstein, B. Gallagher, and C. Faloutsos. DeltaCon: A principled massive-graph similarity function with attribution. *ACM Transactions on Knowledge Discovery from Data*, 2016. DOI: 10.1145/2824443. 4

[127] D. Koutra, H. Tong, and D. Lubensky. Big-align: Fast bipartite graph alignment. In *Proc. of the 14th IEEE International Conference on Data Mining (ICDM)*, Dallas, TX, 2013. DOI: 10.1109/icdm.2013.152. 5, 140

[128] D. Koutra, J. Vogelstein, and C. Faloutsos. DeltaCon: A principled massive-graph similarity function. In *Proc. of the 13th SIAM International Conference on Data Mining (SDM)*, pages 162–170, Austin, TX, 2013. DOI: 10.1137/1.9781611972832.18. 4, 139

[129] L. Kovanen, K. Kaski, J. Kertész, and J. Saramäki. Temporal motifs reveal homophily, gender-specific patterns, and group talk in call sequences. *Proc. of the National Academy of Sciences*, 110(45):18070–18075, 2013. DOI: 10.1073/pnas.1307941110. 94

[130] F. R. Kschischang, B. J. Frey, and H.-A. Loeliger. Factor graphs and the sum-product algorithm. *IEEE Transactions on Information Theory*, 47(2):498–519, 2001. DOI: 10.1109/18.910572. 52

[131] J. Leskovec, D. Chakrabarti, J. M. Kleinberg, and C. Faloutsos. Realistic, mathematically tractable graph generation and evolution, using Kronecker multiplication. In *Proc. of the 9th European Conference on Principles and Practice of Knowledge Discovery in Databases (PKDD)*, pages 133–145, Porto, Portugal, 2005. DOI: 10.1007/11564126_17. 68, 130

[132] J. Leskovec and C. Faloutsos. Sampling from large graphs. In *Proc. of the 12th ACM SIGKDD International Conference on Knowledge Discovery and Data Mining, (KDD'06)*, pages 631–636, New York, ACM, 2006. DOI: 10.1145/1150402.1150479. 46

[133] J. Leskovec, J. Kleinberg, and C. Faloutsos. Graph evolution: Densification and shrinking diameters. *IEEE Transactions on Knowledge and Data Engineering*, vol. 1, March 2007. DOI: 10.1145/1217299.1217301.

[134] J. Leskovec, K. J. Lang, A. Dasgupta, and M. W. Mahoney. Statistical properties of community structure in large social and information networks. In *World Wide Web*, pages 695–704, 2008. DOI: 10.1145/1367497.1367591. 17, 104

[135] J. Leskovec, M. McGlohon, C. Faloutsos, N. S. Glance, and M. Hurst. Patterns of cascading behavior in large blog graphs. In *Proc. of the 7th SIAM International Conference on Data Mining*, Minneapolis, MN, April 26–28, 2007. DOI: 10.1137/1.9781611972771.60. 75

[136] C. Li, J. Han, G. He, X. Jin, Y. Sun, Y. Yu, and T. Wu. Fast computation of SimRank for static and dynamic information networks. In *Proc. of the 13th International Conference on Extending Database Technology, (EDBT'10)*, pages 465–476, New York, ACM, 2010. DOI: 10.1145/1739041.1739098. 50, 140

[137] G. Li, M. Semerci, B. Yener, and M. J. Zaki. Graph classification via topological and label attributes . In *Proc. of the 9th International Workshop on Mining and Learning with Graphs (MLG)*, San Diego, CA, August 2011. 139

[138] M. Li and P. Vitanyi. *An Introduction to Kolmogorov Complexity and its Applications*. Springer, 1993. DOI: 10.1007/978-1-4757-2606-0. 20, 46, 77

[139] Y. Lim, U. Kang, and C. Faloutsos. SlashBurn: Graph compression and mining beyond caveman communities. *IEEE Transactions on Knowledge and Data Engineering*, 26(12):3077–3089, 2014. DOI: 10.1109/tkde.2014.2320716. 17, 26, 29, 30, 46

[140] F. Lin and W. W. Cohen. Semi-supervised classification of network data using very few labels. In *International Conference on Advances in Social Networks Analysis and Mining (ASONAM'10)*, pages 192–199, Odense, Denmark, 2010. DOI: 10.1109/asonam.2010.19. 50

[141] Y. Liu, A. Dighe, T. Safavi, and D. Koutra. A graph summarization: A survey. *CoRR*, abs/1612.04883, 2016. 46, 94

[142] Y. Liu, N. Shah, and D. Koutra. An empirical comparison of the summarization power of graph clustering methods. *arXiv preprint arXiv:1511.06820*, 2015. 47

[143] B. Luo and E. R. Hancock. Iterative procrustes alignment with the EM algorithm. *Image Vision Computing*, 20(5-6):377–396, 2002. DOI: 10.1016/s0262-8856(02)00010-0. 163

[144] A. Maccioni and D. J. Abadi. Scalable pattern matching over compressed graphs via dedensification. pages 1755–1764, ACM, 2016. DOI: 10.1145/2939672.2939856. 46

[145] O. Macindoe and W. Richards. Graph comparison using fine structure analysis. In *International Conference on Privacy, Security, Risk and Trust (SocialCom/PASSAT)*, pages 193–200, 2010. DOI: 10.1109/socialcom.2010.35. 139

[146] P. Mahé and J.-P. Vert. Graph Kernels based on tree patterns for molecules. *Machine Learning*, 75(1):3–35, April 2009. DOI: 10.1007/s10994-008-5086-2. 140

[147] A. S. Maiya and T. Y. Berger-Wolf. Sampling community structure. In *Proc. of the 19th International Conference on World Wide Web (WWW)*, pages 701–710, Raleigh, NC, ACM, 2010. DOI: 10.1145/1772690.1772762. 47

[148] D. M. Malioutov, J. K. Johnson, and A. S. Willsky. Walk-sums and belief propagation in Gaussian graphical models. *Journal of Machine Learning Research*, 7:2031–2064, 2006. 54

[149] H. Maserrat and J. Pei. Neighbor query friendly compression of social networks. In *Proc. of the 16th ACM International Conference on Knowledge Discovery and Data Mining (SIGKDD)*, Washington, DC, 2010. DOI: 10.1145/1835804.1835873. 46

[150] M. McGlohon, S. Bay, M. G. Anderle, D. M. Steier, and C. Faloutsos. SNARE: A link analytic system for graph labeling and risk detection. In *Proc. of the 15th ACM International Conference on Knowledge Discovery and Data Mining (SIGKDD)*, pages 1265–1274, Paris, France, 2009. DOI: 10.1145/1557019.1557155. 52, 53, 141

[151] S. Melnik, H. Garcia-Molina, and E. Rahm. Similarity flooding: A versatile graph matching algorithm and its application to schema matching. In *Proc. of the 18th International Conference on Data Engineering (ICDE)*, San Jose, CA, 2002. DOI: 10.1109/icde.2002.994702. 143, 163

[152] P. Miettinen and J. Vreeken. Model order selection for Boolean matrix factorization. In *Proc. of the 17th ACM International Conference on Knowledge Discovery and Data Mining (SIGKDD)*, pages 51–59, San Diego, CA, 2011. DOI: 10.1145/2020408.2020424. 25

[153] P. Miettinen and J. Vreeken. MDL4BMF: Minimum description length for Boolean matrix factorization. *ACM Transactions on Knowledge Discovery from Data*, 8(4):1–30, 2014. DOI: 10.1145/2601437. 25, 46

[154] R. Milo, S. Itzkovitz, N. Kashtan, R. Levitt, S. Shen-Orr, I. Ayzenshtat, M. Sheffer, and U. Alon. Superfamilies of evolved and designed networks. *Science*, 303(5663):1538–1542, 2004. DOI: 10.1126/science.1089167. 46

[155] R. Milo, S. Shen-Orr, S. Itzkovitz, N. Kashtan, D. Chklovskii, and U. Alon. Network motifs: Simple building blocks of complex networks. *Science*, 298(5594):824–827, 2002. DOI: 10.1126/science.298.5594.824. 46

[156] E. Minkov and W. W. Cohen. Learning to rank typed graph walks: Local and global approaches. In *WebKDD Workshop on Web Mining and Social Network Analysis*, pages 1–8, 2007. DOI: 10.1145/1348549.1348550. 50

[157] A. Narayanan and V. Shmatikov. De-anonymizing social networks. In *Proc. of the 30th IEEE Symposium on Security and Privacy*, pages 173–187, May 2009. DOI: 10.1109/sp.2009.22. 163

[158] S. Navlakha, R. Rastogi, and N. Shrivastava. Graph summarization with bounded error. In *Proc. of the ACM International Conference on Management of Data (SIGMOD)*, pages 419–432, Vancouver, BC, 2008. DOI: 10.1145/1376616.1376661. 46

[159] M. E. Newman. A measure of betweenness centrality based on random walks. *Social Networks*, 27(1):39–54, 2005. DOI: 10.1016/j.socnet.2004.11.009. 141

[160] M. E. J. Newman and M. Girvan. Finding and evaluating community structure in networks. *Physical Review E*, 69(2):026113+, February 2004. DOI: 10.1103/physreve.69.026113. 93

[161] C. C. Noble and D. J. Cook. Graph-based anomaly detection. In *Proc. of the 9th ACM International Conference on Knowledge Discovery and Data Mining (SIGKDD)*, pages 631–636, Washington, DC, ACM, 2003. DOI: 10.1145/956750.956831. 97

[162] OCP. Open connectome project. http://www.openconnectomeproject.org, 2014. 135

[163] J.-P. Onnela, J. Saramäki, J. Hyvönen, G. Szabó, D. Lazer, K. Kaski, J. Kertész, and A.-L. Barabási. Structure and tie strengths in mobile communication networks. *Proc. of the National Academy of Sciences of the USA*, 104(18):7332–6, 2007. DOI: 10.1073/pnas.0610245104. 75

[164] J.-Y. Pan, H.-J. Yang, C. Faloutsos, and P. Duygulu. GCap: Graph-based automatic image captioning. In *4th International Workshop on Multimedia Data and Document Engineering (MDDE)*, page 146, Washington, DC, 2004. DOI: 10.1109/cvpr.2004.353. 50

[165] S. Pandit, D. H. Chau, S. Wang, and C. Faloutsos. NetProbe: A fast and scalable system for fraud detection in online auction networks. In *Proc. of the 16th International Conference on World Wide Web (WWW)*, pages 201–210, Alberta, Canada, 2007. DOI: 10.1145/1242572.1242600. 52, 65

[166] C. H. Papadimitriou and K. Steiglitz. *Combinatorial Optimization: Algorithms and Complexity*. Prentice-Hall, Inc., Upper Saddle River, NJ, 1982. 164

[167] P. Papadimitriou, A. Dasdan, and H. Garcia-Molina. Web graph similarity for anomaly detection. *Journal of Internet Services and Applications*, 1(1):1167, 2008. DOI: 10.1007/s13174-010-0003-x. 116, 117, 138

[168] S. Papadimitriou, J. Sun, C. Faloutsos, and P. S. Yu. Hierarchical, parameter-free community discovery. In *Proc. of the European Conference on Machine Learning and Principles and Practice of Knowledge Discovery in Databases (ECML PKDD)*, Antwerp, Belgium, 2008. DOI: 10.1007/978-3-540-87481-2_12. 47

[169] E. E. Papalexakis, N. D. Sidiropoulos, and R. Bro. From k-means to higher-way co-clustering: Multilinear decomposition with sparse latent factors. *IEEE Transactions on Signal Processing*, 61(2):493–506, 2013. DOI: 10.1109/TSP.2012.2225052. 86

[170] A. Paranjape, A. R. Benson, and J. Leskovec. Motifs in temporal networks. In *Proc. of the 10th ACM International Conference on Web Search and Data Mining, (WSDM'17)*, pages 601–610, New York, ACM, 2017. DOI: 10.1145/3018661.3018731. 94

[171] M. Peabody. Finding groups of graphs in databases. Master's thesis, Drexel University, 2003. 117, 139

[172] J. Pearl. Reverend Bayes on inference engines: A distributed hierarchical approach. In *Proc. of the AAAI National Conference on AI*, pages 133–136, 1982. 50

[173] J. Pearl. *Probabilistic Reasoning in Intelligent Systems: Networks of Plausible Inference*. Morgan Kaufmann, 1988. 51, 52

[174] J. Pei, D. Jiang, and A. Zhang. On mining cross-graph quasi-cliques. In *Proc. of the 11th ACM International Conference on Knowledge Discovery and Data Mining (SIGKDD)*, pages 228–238, Chicago, IL, 2005. DOI: 10.1145/1081870.1081898. 94

[175] B. Perozzi, R. Al-Rfou, and S. Skiena. Deepwalk: Online learning of social representations. In *Proc. of the 20th ACM International Conference on Knowledge Discovery and Data Mining (SIGKDD)*, pages 701–710, New York, ACM, 2014. DOI: 10.1145/2623330.2623732. 168

[176] B. A. Prakash, A. Sridharan, M. Seshadri, S. Machiraju, and C. Faloutsos. Eigen-Spokes: Surprising patterns and scalable community chipping in large graphs. In *Advances in Knowledge Discovery and Data Mining*, pages 435–448, Springer, 2010. DOI: 10.1109/icdmw.2009.103. 17, 21, 26, 83

[177] H. Qiu and E. R. Hancock. Graph matching and clustering using spectral partitions. *IEEE Transactions on Pattern Analysis and Machine Intelligence*, 39(1):22–34, 2006. DOI: 10.1016/j.patcog.2005.06.014. 163

[178] Q. Qu, S. Liu, C. S. Jensen, F. Zhu, and C. Faloutsos. Interestingness-driven diffusion process summarization in dynamic networks. In *Proc. of the European Conference on Machine Learning and Principles and Practice of Knowledge Discovery in Databases (ECML PKDD)*, pages 597–613, Nancy, France, 2014. DOI: 10.1007/978-3-662-44851-9_38. 95

[179] D. Rafiei and S. Curial. Effectively visualizing large networks through sampling. In *16th IEEE Visualization Conference (VIS)*, page 48, Minneapolis, MN, 2005. DOI: 10.1109/visual.2005.1532819. 47

[180] J. Ramon and T. Gärtner. Expressivity vs. efficiency of graph Kernels. In *Proc. of the 1st International Workshop on Mining Graphs, Trees and Sequences*, pages 65–74, 2003. 140

[181] S. Ranshous, S. Shen, D. Koutra, S. Harenberg, C. Faloutsos, and N. F. Samatova. Graph-based anomaly detection and description: A survey. *WIREs Computational Statistics*, January (accepted) 2015. 141

[182] K. Riesen and H. Bunke. Approximate graph edit distance computation by means of bipartite graph matching. *Image and Vision Computing*, 27(7):950–959, 2009. DOI: 10.1016/j.imavis.2008.04.004. 163

[183] J. Rissanen. Modeling by shortest data description. *Automatica*, 14(1):465–471, 1978. DOI: 10.1016/0005-1098(78)90005-5. 18, 46

[184] J. Rissanen. A universal prior for integers and estimation by minimum description length. *The Annals of Statistics*, 11(2):416–431, 1983. DOI: 10.1214/aos/1176346150. 20, 23, 80

[185] W. G. Roncal, Z. H. Koterba, D. Mhembere, D. Kleissas, J. T. Vogelstein, R. C. Burns, A. R. Bowles, D. K. Donavos, S. Ryman, R. E. Jung, L. Wu, V. D. Calhoun, and R. J. Vogelstein. MIGRAINE: MRI graph reliability analysis and inference for connectomics. *IEEE Global Conference on Signal and Information Processing (GlobalSIP)*, 2013. DOI: 10.1109/globalsip.2013.6736878. 135

[186] M. Rosvall and C. T. Bergstrom. An information-theoretic framework for resolving community structure in complex networks. *Proc. of the National Academy of Sciences*, 104(18):7327–7331, 2007. DOI: 10.1073/pnas.0611034104. 47

[187] C. Schellewald and C. Schnörr. Probabilistic subgraph matching based on convex relaxation. In *Energy Minimization Methods in Computer Vision and Pattern Recognition*, pages 171–186, Springer, 2005. DOI: 10.1007/11585978_12. 163

[188] N. Shah, A. Beutel, B. Gallagher, and C. Faloutsos. Spotting suspicious link behavior with fBox: An adversarial perspective. In *Proc. of the 14th IEEE International Conference on Data Mining (ICDM)*, Shenzhen, China, IEEE, 2014. DOI: 10.1109/icdm.2014.36. 93

[189] N. Shah, D. Koutra, T. Zou, B. Gallagher, and C. Faloutsos. TimeCrunch: Interpretable dynamic graph summarization. In *Proc. of the 21st ACM International Conference on Knowledge Discovery and Data Mining (SIGKDD)*, Sydney, Australia, 2015. DOI: 10.1145/2783258.2783321. 4, 75

[190] N. Shervashidze and K. Borgwardt. Fast subtree Kernels on graphs. In *23rd Annual Conference on Neural Information Processing Systems (NIPS)*, pages 1660–1668, Vancouver, British Columbia, 2009. 140

[191] N. Shervashidze, P. Schweitzer, E. J. van Leeuwen, K. Mehlhorn, and K. M. Borgwardt. Weisfeiler-Lehman graph Kernels. *Journal of Machine Learning Research*, 12:2539–2561, November 2011. 140

[192] N. Shervashidze, S. V. N. Vishwanathan, T. Petri, K. Mehlhorn, and K. Borgwardt. Efficient graphlet Kernels for large graph comparison. In *Proc. of the 12th International Conference on Artificial Intelligence and Statistics (AISTATS)*, volume 5, pages 488–495, Journal of Machine Learning Research, 2009. 140

[193] J. Shetty and J. Adibi. The Enron e-mail dataset database schema and brief statistical report. *Information Sciences Institute Technical Report*, University of Southern California, 2004.

[194] L. Shi, H. Tong, J. Tang, and C. Lin. Vegas: Visual influence graph summarization on citation networks. *IEEE Transactions on Knowledge and Data Engineering*, 27(12):3417–3431, 2015. DOI: 10.1109/tkde.2015.2453957. 95

[195] B. Shneiderman. Extreme visualization: Squeezing a billion records into a million pixels. In *Proc. of the ACM International Conference on Management of Data (SIGMOD)*, Vancouver, BC, 2008. DOI: 10.1145/1376616.1376618. 47

[196] R. Singh, J. Xu, and B. Berger. Pairwise global alignment of protein interaction networks by matching neighborhood topology. In *Proc. of the 11th Annual International Conference on Computational Molecular Biology (RECOMB)*, pages 16–31, San Francisco, CA, 2007. DOI: 10.1007/978-3-540-71681-5_2. 163, 165

[197] A. Smalter, J. Huan, and G. Lushington. GPM: A graph pattern matching Kernel with diffusion for chemical compound classification. In *Proc. of the IEEE International Symposium on Bioinformatics and Bioengineering (BIBE)*, 2008. DOI: 10.1109/bibe.2008.4696654. 143, 163

[198] SNAP. http://snap.stanford.edu/data/index.html#web

[199] S. Soundarajan, T. Eliassi-Rad, and B. Gallagher. A guide to selecting a network similarity method. In *Proc. of the 14th SIAM International Conference on Data Mining (SDM)*, pages 1037–1045, Philadelphia, PA, 2014. DOI: 10.1137/1.9781611973440.118. 140

[200] K. Sricharan and K. Das. Localizing anomalous changes in time-evolving graphs. In *Proc. of the ACM International Conference on Management of Data (SIGMOD)*, pages 1347–1358, Snowbird, UT, ACM, 2014. DOI: 10.1145/2588555.2612184. 123, 128, 141

[201] J. Sun, C. Faloutsos, S. Papadimitriou, and P. S. Yu. GraphScope: Parameter-free mining of large time-evolving graphs. In *Proc. of the 13th ACM International Conference on Knowledge Discovery and Data Mining (SIGKDD)*, pages 687–696, San Jose, CA, ACM, 2007. DOI: 10.1145/1281192.1281266. 93

[202] J. Tang, M. Qu, M. Wang, M. Zhang, J. Yan, and Q. Mei. Line: Large-scale information network embedding. In *Proc. of the 24th International Conference on World Wide Web (WWW)*, pages 1067–1077, Florence, Italy, 2015. DOI: 10.1145/2736277.2741093. 168

[203] J. Tang, J. Sun, C. Wang, and Z. Yang. Social influence analysis in large-scale networks. In *KDD*, pages 807–816, ACM, 2009. DOI: 10.1145/1557019.1557108. 75

[204] N. Tang, Q. Chen, and P. Mitra. Graph stream summarization: From big bang to big crunch, pages 1481–1496, 2016. DOI: 10.1145/2882903.2915223. 95

[205] N. Tatti and J. Vreeken. The long and the short of it: Summarizing event sequences with serial episodes. In *Proc. of the 18th ACM International Conference on Knowledge Discovery and Data Mining (SIGKDD)*, Beijing, China, ACM, 2012. DOI: 10.1145/2339530.2339606. 45

[206] S. L. Tauro, C. Palmer, G. Siganos, and M. Faloutsos. A simple conceptual model for the internet topology. *IEEE Global Telecommunications Conference (GLOBECOM'01)*, 2001. DOI: 10.1109/glocom.2001.965863. 17, 21

[207] Y. Tian, R. A. Hankins, and J. M. Patel. Efficient aggregation for graph summarization. In *Proc. of the ACM International Conference on Management of Data (SIGMOD)*, pages 567–580, Vancouver, BC, 2008. DOI: 10.1145/1376616.1376675. 46

[208] H. Toivonen, F. Zhou, A. Hartikainen, and A. Hinkka. Compression of weighted graphs. In *Proc. of the 17th ACM International Conference on Knowledge Discovery and Data Mining (SIGKDD)*, pages 965–973, San Diego, CA, 2011. DOI: 10.1145/2020408.2020566. 46, 94

[209] H. Tong, C. Faloutsos, and J.-Y. Pan. Fast random walk with restart and its applications. In *Proc. of the 6th IEEE International Conference on Data Mining (ICDM)*, pages 613–622, Hong Kong, China, 2006. DOI: 10.1109/icdm.2006.70. 50, 55

[210] H. Tong, B. A. Prakash, T. Eliassi-Rad, M. Faloutsos, and C. Faloutsos. Gelling, and melting, large graphs by edge manipulation. In *Proc. of the 21st ACM Conference on Information and Knowledge Management (CIKM)*, pages 245–254, Maui, Hawaii, ACM, 2012. DOI: 10.1145/2396761.2396795. 141

[211] J. Ugander, L. Backstrom, and J. Kleinberg. Subgraph frequencies: Mapping the empirical and extremal geography of large graph collections. In *Proc. of the 22nd International Conference on World Wide Web*, pages 1307–1318, ACM, 2013. DOI: 10.1145/2488388.2488502. 46

[212] S. Umeyama. An Eigen decomposition approach to weighted graph matching problems. *IEEE Transactions on Pattern Analysis and Machine Intelligence*, 10(5):695–703, 1988. DOI: 10.1109/34.6778. 144, 145, 150, 157, 164

[213] F. van Ham, H.-J. Schulz, and J. M. Dimicco. Honeycomb: Visual analysis of large scale social networks. In *Human-Computer Interaction—INTERACT*, volume 5727 of *Lecture Notes in Computer Science*, pages 429–442, Springer Berlin Heidelberg, 2009. DOI: 10.1007/978-3-642-03658-3_47. 139

[214] S. V. N. Vishwanathan, N. N. Schraudolph, R. I. Kondor, and K. M. Borgwardt. Graph Kernels. *Journal of Machine Learning Research*, 11:1201–1242, 2010. 140

[215] B. Viswanath, A. Mislove, M. Cha, and K. P. Gummadi. On the evolution of user interaction in Facebook. In *Proc. of the 2nd ACM SIGCOMM Workshop on Social Networks (WOSN)*, Barcelona, Spain, August 2009. DOI: 10.1145/1592665.1592675. 161

[216] J. T. Vogelstein, J. M. Conroy, L. J. Podrazik, S. G. Kratzer, D. E. Fishkind, R. J. Vogelstein, and C. E. Priebe. Fast inexact graph matching with applications in statistical connectomics. *CoRR*, abs/1112.5507, 2011. 144, 145, 164

[217] J. Vreeken, M. van Leeuwen, and A. Siebes. KRIMP: Mining itemsets that compress. *Data Mining and Knowledge Discovery*, 23(1):169–214, 2011. DOI: 10.1007/s10618-010-0202-x. 45

[218] D. Wang, P. Cui, and W. Zhu. Structural deep network embedding. *KDD*, 2016. DOI: 10.1145/2939672.2939753. 168

[219] Y. Wang, S. Parthasarathy, and S. Tatikonda. Locality sensitive outlier detection: A ranking driven approach. In *Proc. of the 27th International Conference on Data Engineering (ICDE)*, pages 410–421, Hannover, Germany, 2011. DOI: 10.1109/icde.2011.5767852. 97

[220] D. J. Watts. *Small Worlds: The Dynamics of Networks between Order and Randomness*. Princeton University Press, 1999. DOI: 10.1063/1.1333299. 17, 88

[221] Y. Weiss. Correctness of local probability propagation in graphical models with loops. *Neural Computation*, 12(1):1–41, 2000. DOI: 10.1162/089976600300015880. 54

[222] R. C. Wilson and P. Zhu. A study of graph spectra for comparing graphs and trees. *Journal of Pattern Recognition*, 41(9):2833–2841, 2008. DOI: 10.1016/j.patcog.2008.03.011. 117, 139

[223] K. S. Xu, M. Kliger, and A. O. Hero III. Tracking communities in dynamic social networks. In *Proc. of the 4th International Conference on Social Computing, Behavioral-Cultural Modeling, and Prediction (SBP'11)*, pages 219–226, Springer, 2011. DOI: 10.1007/978-3-642-19656-0_32. 94

[224] Yahoo! Webscope. `webscope.sandbox.yahoo.com`

[225] X. Yan and J. Han. gSpan: Graph-based substructure pattern mining. In *IEEE International Conference on Data Mining*, Los Alamitos, CA, IEEE Computer Society Press, 2002. DOI: 10.1109/icdm.2002.1184038. 45

[226] Ö. N. Yaveroğlu, N. Malod-Dognin, D. Davis, Z. Levnajic, V. Janjic, R. Karapandza, A. Stojmirovic, and N. Pržulj. Revealing the hidden language of complex networks. *Scientific Reports*, 4, 2014. DOI: 10.1038/srep04547. 46

[227] Ö. N. Yaveroğlu, N. Malod-Dognin, D. Davis, Z. Levnajić, V. Janjic, R. Karapandza, A. Stojmirovic, and N. Pržulj. Revealing the hidden language of complex networks. *Scientific Reports*, 4, 2014. DOI: 10.1038/srep04547. 140

[228] J. S. Yedidia, W. T. Freeman, and Y. Weiss. Understanding belief propagation and its generalizations. In *Exploring Artificial Intelligence in the New Millennium*, pages 239–269, 2003. 52, 140, 164

[229] J. S. Yedidia, W. T. Freeman, and Y. Weiss. Constructing free-energy approximations and generalized belief propagation algorithms. *IEEE Transactions on Information Theory*, 51(7):2282–2312, 2005. DOI: 10.1109/tit.2005.850085. 52

[230] W. Yu, X. Lin, W. Zhang, L. Chang, and J. Pei. More is simpler: Effectively and efficiently assessing node-pair similarities based on hyperlinks. *Proc. of the VLDB Endowment*, 7(1):13–24, 2013. DOI: 10.14778/2732219.2732221. 50, 140

[231] L. Zager and G. Verghese. Graph similarity scoring and matching. *Applied Mathematics Letters*, 21(1):86–94, 2008. DOI: 10.1016/j.aml.2007.01.006. 163

[232] M. Zaslavskiy, F. Bach, and J.-P. Vert. A path following algorithm for the graph matching problem. *IEEE Transactions on Pattern Analysis and Machine Intelligence*, 31(12):2227–2242, December 2009. DOI: 10.1109/tpami.2008.245. 143, 144, 145, 164

[233] N. Zhang, Y. Tian, and J. M. Patel. Discovery-driven graph summarization. In *Proc. of the 26th International Conference on Data Engineering (ICDE)*, pages 880–891, Long Beach, CA, 2010. DOI: 10.1109/icde.2010.5447830. 46

[234] Q. Zhao, Y. Tian, Q. He, N. Oliver, R. Jin, and W.-C. Lee. Communication motifs: A tool to characterize social communications. In *Proc. of the 19th ACM International Conference on Information and Knowledge Management*, pages 1645–1648, ACM, 2010. DOI: 10.1145/1871437.1871694. 94

[235] X. Zhu. Semi-supervised learning literature survey, 2006. 50, 51, 53, 56

[236] X. Zhu, Z. Ghahramani, and J. Lafferty. Semi-supervised learning using Gaussian fields and harmonic functions. In *Proc. of the 20th International Conference on Machine Learning (ICML)*, pages 912–919, Washington, DC, 2003. 56

Authors' Biographies

DANAI KOUTRA

Danai Koutra is an Assistant Professor in Computer Science and Engineering at University of Michigan, Ann Arbor. Her research interests include large-scale graph mining, graph similarity and matching, graph summarization, and anomaly detection. Danai's research has been applied mainly to social, collaboration, and web networks, as well as brain connectivity graphs. She holds one "rate-1" patent and has six (pending) patents on bipartite graph alignment. Danai won the 2016 ACM SIGKDD Dissertation award, and an honorable mention for the SCS Doctoral Dissertation Award (CMU). She has multiple papers in top data mining conferences, including two award-winning papers, she has given three tutorials, and her work has been covered by the popular press, such as the MIT Technology Review. She has worked at IBM Watson, Microsoft Research, and Technicolor. She earned her Ph.D. and M.S. in Computer Science from CMU in 2015 and her diploma in Electrical and Computer Engineering at the National Technical University of Athens in 2010.

CHRISTOS FALOUTSOS

 **Christos Faloutsos** is a Professor at Carnegie Mellon University. He has received the Presidential Young Investigator Award by the National Science Foundation (1989), the Research Contributions Award in ICDM 2006, the SIGKDD Innovations Award (2010), 24 "best paper" awards (including 5 "test of time" awards), and 4 teaching awards. Six of his advisees have attracted KDD or SCS dissertation awards, He is an ACM Fellow, he has served as a member of the executive committee of SIGKDD; he has published over 350 refereed articles, 17 book chapters, and 2 monographs. He holds seven patents (and 2 pending), and he has given over 40 tutorials and over 20 invited distinguished lectures. His research interests include large-scale data mining with emphasis on graphs and time sequences; anomaly detection, tensors, and fractals.

Printed in the United States
by Baker & Taylor Publisher Services